ELEMENTS
OF
STABLE MANAGEMENT

Also by Carol R. Melcher:

The Beginner's Guide to Horses: Buying, Equipping, and Stabling

ELEMENTS OF STABLE MANAGEMENT

An Illustrated Guide

Carol R. Melcher

SOUTH BRUNSWICK AND NEW YORK: A. S. BARNES AND COMPANY

LONDON: THOMAS YOSELOFF LTD

A. S. Barnes and Co., Inc.
Cranbury, New Jersey 08512

Thomas Yoseloff Ltd
108 New Bond Street
London W1Y OQX, England

Library of Congress Cataloging in Publication Data

Melcher, Carol R
    Elements of stable management.

    Includes index.
    1. Stables.   2. Horses.   I. Title.
SF318.M44   636.1'08'3   74-9291
ISBN 0-498-01517-3

PRINTED IN THE UNITED STATES OF AMERICA

CONTENTS

PREFACE

One of the most important yet one of the least covered subjects in the field of horsemanship literature today is that of stable management. There are many books on the market today that deal with riding instruction and horse training, but there are few that deal with and are totally committed to stable management. To me, stable management is as important, or in some instances more important, than the training of horse and rider.

Too many people fancy themselves horsemen who have absolutely no idea, whatsoever, of how to properly care for a horse and stable.

A horse can be trained to the limit of capacity, making him what is known as a "push button" horse, but if the management of the stable is poor then he will, more often than not, be able to give less than his utmost performance.

Stable management means more than just feeding, watering, and cleaning stalls. It means the housing of the horse, his total comfort, well being, care, grooming, and keeping of his tack, as well as the quality of his feed, bedding, and the construction of his house. It means knowing how and when to pasture him and how to use all of the right things in the right places and at the right times. It also means knowing when and if to have him shod, having a good worming and pest control schedule, as well as free and controlled exercise times.

Many horses today are still kept with little regard to their comfort. Whether they are hot or cold, in a draft or stifling for lack of fresh air is disregarded. Many owners feel that all that is needed is a stall, feed, and water. They disregard the stall size, its construction, its weather tightness, its airiness, and whether it is light enough or even clean enough.

This book was written with the thought of producing a work that would not just tell the owner what to do for his horse, but would explain how to do for his horse.

Included in this book are chapters dealing with, first of all, the stable, its construction and the various materials and designs for basic comfort. Included therein is a section on stalls: how to build one from scratch or how to convert a stall that is already in place in the barn; the best and worst kind of floors for a stall and why each has its merits and disadvantages. Included is the equipment needed in each stall and the best and worst kind of each. Bedding is discussed, as is feed, and a full explanation of what to look for and avoid when buying both, and why. The advantages of having a cross-tie area or wash rack is discussed, as is that of having a large box stall with cross-ties for isolation and/or foaling purposes.

Storage of feed and bedding and containers for both is dealt with as is the storage of saddles, bridles, and all other loose stable equipment.

Pasture is talked about, along with how to get your horse ready to turn out to grass. The advantages of having a pasture with running water, rather than a still pond, is explained, as is the reason that you should have your horse innoculated against disease, for his own protection and for the protection of others, including humans.

To shoe or not to shoe, when, why, and why not are discussed, as are the advantages and disadvantages of both. Also there is a brief talk on trailering, or vanning your horse in a truck and the for and against for both methods.

Contained in this book are photos of good and bad stables as well as good and bad pasture as well as containers for feed and good and bad tackroom set up. There is also a discussion of conditions for keeping the loose stable equipment, such as shovels and pitchforks.

Stable management should not be a sometimes thing. It should be a full time project, so that it becomes ingrained, a part of the horseman. Good horse care is an absolute must for anyone who expects the best performance from his horse. Granted, stable work is just that, work, but it can also be pleasurable, in that a true horseman derives satisfaction by doing things that he knows are going to bring comfort and pleasure to his horse. Sometimes grooming a horse can be as much of a pleasure as riding, and watching a horse enjoy his feed is satisfying in itself.

Stable management, the management of a stable, is the ultimate concern for every true horseman.

ACKNOWLEDGMENTS

The author gratefully thanks the following stables for cooperation in photographing their property, animals, and facilities:

Northhill Farm, R.D. #1, Oley, Pa.—Owner, Mrs. Howard C. Paddock.

Furnace Hill Stables, R.D. #1, Fleetwood, Pa.—Owner-manager, Mr. Richard T. Curtin.

Long "D" Ranch, R.D. #1, Reinholds, Pa.—Owner, Mr. Richard DeLong.

All other photos, not from the above sources, were drawn from the author's own photo files.

ELEMENTS
OF
STABLE MANAGEMENT

INVOCATION

When cold and wet, please rub me dry,
And do not beat me when I shy;
Give twice a week a hot bran mash,
With corn and oats, and salt, a dash;
Ten pounds each day of hay that's free
From dust, all you should give to me;
Feed twice a week, instead of oats,
A pair of carrots, 'twill shine my coat;
When hot, don't give me drink or grain;
When cold, don't stand me in the rain;
Batten my stable warm and tight,
And see that it's kept clean and light.
In winter, blanket close and bed me deep,
And you'll find I'll pay you for my keep.

1

STABLE CONSTRUCTION

IN THIS CHAPTER I am going to deal with as many stable types as possible—from the small, one-horse stable, to the large, racing or show stables. Included are illustrations of each type and its description.

To begin with, let's discuss stable construction itself. As with any building, the stable may or may not be properly built. Too often you see horses housed in sheds, lean-tos, and falling down, delapidated buildings. If you intend to have a horse, you should be willing to house him properly. A good stable will always have a footing that is placed below the frost line. This keeps the foundation from heaving up during freezes and thaws of the ground. The best foundations are made of cinder or cement block. The barn itself can be of many different types of material: cinder block, wood, brick, poles, or corrugated, structural steel. Depending on your particular region or area, one or more of these materials may or may not be suitable for stable construction.

If you are building a one-horse barn you will need room for one stall, a tack room, a feed storage area, and a grooming area. If you are building for more than one horse you may want to add an indoor riding area or a lounge, if you intend to board other horses.

Regardless of how many or few horses you intend to keep, you want to be sure to house them properly.

One of the first things to consider when building a stable is to decide what type floor you are going to have. The best floor is one of dirt over a pit filled with stones. The ideal arrangement is to dig down about eighteen inches in the entire area where the stall floor is going to be located. After the excavation has been completed, fill the pit in with stones that range from six inches to one foot across, putting the smaller stones on top. Make the stones fit fairly well together in order to produce a solid base. Fill the pit with stones to a depth of about five to six inches, then cover with dirt. After the pit is filled, walk the dirt down until it is flat, smooth, and firm, adding more dirt as you go, if need be. This flooring-over-pit arrangement provides maximum drainage of the stall and keeps the bedding and the horses feet much drier. Drainage is one of the most often ignored areas of stable management and one of the worst problems to both horse and manager, if not properly taken care of. Now and then, over a period of perhaps a year, you will have to add some more dirt to the stall floor as the old dirt wears away from the horse walking about on it. Also, some dirt will be removed when you muck out the stall each week. Be sure to see, at each weekly total clean out of the stall, that none of the rocks in the pit have worked to the top and are protruding. This not only makes it uncomfortable for the horse, but dangerous. If he steps on the rock often enough he may bruise the sole or frog of his hoofs. Or, if he lies on it, he may injure himself elsewhere.

The next acceptable flooring is the solid dirt one. While this flooring does not permit the stall wastes to drain away as rapidly as does the dirt-over-pit arrangement, it is suitable, as long as it is kept clean.

Wooden flooring, such as in upper floors of a barn, is suitable only if it is inspected carefully and often—and only if it does not drain into a stall below it. Wood is very susceptible to rot, and the ammonia in the horse's urine increases the rotting action. Wood also holds dampness, and a damp stall is one of the worst hazards of stable management. Also, if your horse is inclined to paw a lot, he may paw a hole in the wood if he does his pawing in one spot. Or he may split the wood and pick up a bad splinter or shaft of wood in his foot. Wooden floors also tend to be slippery when wet or damp with manure and urine.

The least acceptable, and I mean least, is a concrete floor. Some people think that concrete is good because you can wash it down and disinfect the stall. This is true; however, there is no drainage with concrete, unless each stall is made with a gutter out to the alley way. (Stalls built in this fashion are seldom seen in the United States, but are fairly common in Europe and Great Britain.)

Concrete is also cold. It does not maintain any of the horse's body heat when used as flooring, nor is it good for him to stand on it for long periods of time. Concrete floors are hard on the feet and legs, whether they belong to beast or man.

As you can see, the ideal stall floor is a natural one—one of dirt or earth. The closer you keep your horse to nature, the better off he will be.

An excellent example of large, sliding doors at the end of a barn.

Another vexing problem with stalls is proper ventilation. It is necessary to keep the stable well aired, yet keep the horses out of a direct draft. The best idea is for each stall to have its own window, or half door, to the outside and to have a barn with double doors at each end.

This way, any number of windows may be opened or shut, as need be. If it is hot, as on a still summer day, all the windows may be opened and the double doors at the ends of the barn as well. If the wind is from one particular direction, the windows on that side of the barn may be closed, while the rest remain open. This allows for free circulation of air without draft.

Windows that tilt out are good for stables. This permits the air to sweep in and upward, passing above the horse, instead of hitting him

This is a good example of a tilt window and the guard bars used to keep the horse from putting his nose through the glass.

directly. This keeps the stable air moving freely, yet does not touch the horse.

Windows that are too high do no good. The air merely comes in one window, passes along the roof line or ridge pole of the building, and passes out the window directly opposite. The windows should be set at just about the same height that the bars of the stall begin. This is about a foot higher than normal house windows are set or at about withers height. This permits free circulation of air to clean the stale air out of the barn with the least amount of draft. All windows should be either barred or heavily screened on the inside, to prevent the horse putting his head through the glass.

Another little discussed problem in stables is the lighting. Many, many stables are poorly lit. Proper stable lighting is that which allows a person to see a horse well enough to immediatley detect any cuts, scars, or abnomalities, as well as he would be able to do in the daytime. Poor lighting is asking for trouble. If a horse is kept in the stable for any length of time, and the lighting is poor (this goes for daytime light as well), his sight may become impaired. If you cannot spot cuts or abnormalities right away, because of bad lighting, it could cost you the life of your horse.

Outside view of a well barred stall window. It may be opened from either the top or the bottom, by raising or lowering the lower or upper pane of glass.

This large barn has a number of skylights that provide a good bit of daylight for the barn.

Poor lighting also encourages rodents to prowl around, looking for food in the half light.

Skylights are excellent for daytime additional lighting, and windows in each stall help immensely.

Doors that open to the outside at each stall may have the top half open on good days, to allow the horse to put his head outside if he wishes. This is a good type of stall to use if you have a horse that is confined to barn for a long period of time, such as with an injury.

Lights for the barn should be florescent wherever possible. Florescent lights cost less to operate than incandescent bulbs, they burn longer without getting nearly as hot, and they shed more light over a greater amount of space. Lights in stalls may be incandescent bulbs, if they are covered with a wire shield. This shield protects both the horse and the bulb, should the horse rear in his stall and strike the fixture. If the bulbs are not guarded, the horse may get cut and/or badly damage the electrical fixture. All aisle and wash rack or saddling area lights should be florescent, as should lights for an indoor riding arena. Florescent lights also cast less shadow than do incandescent bulbs and this is often a helping factor when training a young or spooky horse inside. Economy, more light for less money and less shadows, all add up to better lighting for the stable.

Now, what about the aisles of the barn? Unlike the stalls, which should be dirt, one of the best things for barn aisles is cold rolled asphalt. This is laid directly over a smoothed, prepared dirt floor. It is rolled with a roller by the firm that installs it and it is low in temperature when put

Half doors, like this, are excellent for all stalls giving directly to the outside.

down. It is often used to patch parking lots and in some parts of the country goes under the name of Cold Patch. Cold rolled asphalt is better than hot because it does not bind together as closely and is not as slippery as are many asphalts used for highway and driveway construction. The cold asphalt provides firm footing for the horse, his shoes will not slip on it as they will on the tighter knit asphalts, it is easily cleaned and may be scrubbed with a brush and disinfectant if need be. Just make sure that it is rolled flush with the entrance to the stall and does not produce a ledge that the horse may trip or fall over. The stall floor may be built up to meet it if necessary. Keep check on the condition of the edges of the aisle. If the level between the aisle and the stall floor changes, fill in and pack the dirt down well until a flush level is reached between the two.

Deeply worn ruts and protruding rocks are a constant problem in dirt-aisled barns.

Another common aisle floor is concrete. This is acceptable from a cleanliness standpoint, since concrete will take nearly anything, from manure and urine to cleaners, disinfectants, etc. However, concrete is poor from a safety standpoint. It is slippery against metal-shod feet and a horse can go down quite unexpectedly in a concrete aisle. If your barn has a concrete aisle, be sure to walk your horse slowly in the barn and enter and exit with caution. The one advantage that concrete does have for aisles is that concrete aprons can be run into each stall, thus making a flat entrance and exit from the stall. Dirt may be laid directly on top of the apron or may be swept back from it in order to keep the stall entrance free of stable litter. However, the danger of slipping in the concrete aisle is still a very real one.

Dirt aisles are hard to maintain. A stable that is used a lot, with quite a bit of four and two footed traffic going in and out, soon develops worn paths and ruts.

Constant maintenance and filling and packing is necessary to keep the floor in good shape. Too, there are often rocks that protrude from dirt floors, and they present a definite tripping hazard, both for you and for the horse. Dirt aisles provide the best footing of any aisles, but are the hardest to maintain.

Now, let's discuss what you use to build a stable. Cinder block or stone makes a good stable, though stone is one of the most expensive materials you can use in building a barn.

Brick is also a good material with which to build a stable that is going to last. Cinder block, brick, and stone all build stables that are more fireproof than any others. All three help contain body heat, since they are firm and tight in construction. However, they are fairly expensive to construct. This is one of the reasons that wood is most often used to build barns and stables. Wood also holds body heat, though not as well because it cannot be built or constructed as tightly as barns made of brick or stone. However, wood will absorb the heat of the sun on sunny winter days whereas stone and brick will not.

Of course, wood is one of the least fireproof materials that can be used to build with. If you house your horses in a wooden barn or stable, then by all means you should not keep feed or grain stored overhead and not too large a quantity of either in the lower barn. Fire spreads quickly in any barn, but it spreads faster through a wooden one. This is the primary reason for having a separate building for feed and bedding storage.

One alternative is to build with contoured metal. While metal barns do not hold much body heat, some of them are designed so that the body heat, while it rises, remains in the building at least. Quonset hut style

This barn has stone end walls and a stone front under the wooden upper floors. Though expensive to construct, these stone barns are among the tightest and most draft free.

barns, with dome shaped roofs are ideal for keeping body heat down near the horses where it is most needed in winter. The arch shaped roof holds the heat along the walls, thus acting as a nearly natural insulation for the animals. The structural steel buildings, with straight walls, however, while alright for indoor riding arenas, do leave much to be desired in the way of stabling due to the poor quality of body heat maintenance. While they are easy to maintain, in the way of not needing paint, etc., they are not good in cold climates in that they do not retain body heat for the animals. Often, on a very cold winter day, it is colder in the barn than outside, due to its usual high ceiling and lack of insulation.

Stall construction is another factor to think about when building a

This is a fine example of a two story, frame barn with a stone enclosed paddock area.

barn or remodeling one that you might already have. To begin with you should see that the stall is big enough or that you will make it big enough. A minimum of ten feet square is required and twelve feet square is even better. A horse must be able to move around with ease and comfort while stabled, to remain healthy and sound. The stalls should be constructed of oak boards spaced approximately one to two inches apart. They should be about one inch thick. A horse can put out a lot of power, even with a kick that has been thrown in play or boredom. If you don't want to spend all your time replacing stall boards, it is best to build right in the beginning. A feed tub or box should be installed in a corner of the stall where the horse is least likely to become caught by it or least likely

This barn is built of structural steel, is easily ventilated, and has large escape doors.

to knock it off the wall. A poor place to put it is right by the door where he walks in and out. I advocate using the plastic tubs rather than building a wooden feed box or hay manger. Too often feed becomes dampened by the horse's breath and saliva and clings to the bottom of the wooden boxes and mangers. If this feed is allowed to dry and is not removed, it can mold or contaminate other feed that is put in on top of it. It can also sour and produce bad odors in the stable. Plastic tubs, on the other hand, can be easily removed and hosed and scrubbed out. Also, it is much easier to see any feed that may have been left in the tub and remove it before it has a chance to dry or collect. Too, if your horse should get sick or catch a cold, the plastic tubs can be removed and disinfected, whereas the wooden mangers and boxes cannot.

Wooden boxes pose another problem. Small bits of grain can get stuck under the lip of the manger, if it is the old type, and they can dry there. Sooner or later, the scent of the old, dried oats will bring rats and mice. The less bait there is in the stable, the less trouble will be had with vermin.

Wooden hay mangers are another thing found in most stalls or else, the curse of the horse, the hay rack. While hay racks do keep the hay compacted and they do make the horse eat more slowly, thus lessening the chances of his getting colic from bolting his food, they are a plague when it comes to the horse's general health. Seeds, dust, and small bits of loose hay fall down with each mouthful the horse pulls from the pat of hay that is in the rack. These bits and pieces fall directly onto the

A good example of a very well constructed stall, made of solid-oak boards.

This is a good set up for feeding and watering. The water bucket is first, then the grain tub, and then the hay manger.

horse's face, often into his eyes and up his nose, and are sometimes inhaled into the lungs. Small insects as well may be in the hay, and if they should happen to fall into the eyes, they may lodge there.

A better idea is the wooden manger, though that is not entirely satisfactory. Wooden mangers for hay hold the pat so that the horse must put his head down into it to eat his hay, and while downward is the proper position for a horse to feed, it is bad for him to do so in a confined space. Again, if the hay happens to be even a bit dusty, the dust stirred up by him shaking out the pat of hay may easily be inhaled into the lungs, thus setting up the ideal conditions for a lung infection or coughing problem. Some wooden mangers have holes drilled into the bottoms and sides so that all loose dirt and bits of hay can be brushed out completely, and this is a good idea. This way the bits do not collect in the bottom of the manger and mold. The holes should be a bit bigger than a quarter and spaced about two inches to four inches apart. As stated before, this is a better method than the hay rack, but it is not ideal. The ideal method of feeding hay is to feed it on the floor of the stall, right next to, but not underneath, the oats tub. Sure, some bit of hay will be wasted, but if you feed a proper amount and not too much, the horse will eat the amount of hay you have given him and leave very little as waste. Feeding in this manner puts the horse in the natural grazing position, the dust settles immediately to the floor, and if you have kept his bed clean as you

This is an example of hardened, matted hay bits that have been allowed to collect in the bottom of a manger. It is nearly as hard as concrete.

should, he will not pick up any foreign matter, whether in his eyes, nose, or mouth, when he feeds in this manner.

Now, what about water? Well, a horse should have water in front of him at all times. Horses that cannot drink when they would, often develop bowel troubles, such as constipation or impactions. Horses that cannot drink whenever they want or do not have water at all times often hurriedly drink their fill when given water and then colic, especially if they are fed right away.

A large hook that has had the protruding hook or point filed off is ideal when fastened to the wall of the stall. It holds a bucket of water securely and the bucket is easily removed for refilling and cleaning. A

Example of wooden feed box for grain. Grain can easily dry and lodge in the corners and along the front lip of the box.

An example of a wooden manger. Note the holes drilled into the front panel of the manger.

wooden shelf with a ledge is also a good idea for holding a bucket and can easily be built into the stall. Place a rubber door guard around the bucket and clip both ends of it to the wall in order to keep the bucket in place.

I think a bucket in the stall is better than the automatic waterers, only in that too often the automatic waterers can get jammed, either open or closed. If jammed open they will flood the stall. If jammed closed the horse goes thirsty. Too, a bucket is easier and faster to clean than is an automatic waterer. If you use a plastic bucket for water in winter, there is also no chance of the horse getting his lips or tongue frozen to it as there is with a metal bucket or waterer.

Though not the best setup for hay and water, the wooden shelf holding the water bucket is ideal. The bucket can easily be placed into and removed from this holder.

Place the bucket so that the horse can put his nose in it with comfort. About chest height is good—high enough to keep his feet out of it yet low enough for ease of drinking. This should be placed next to the grain tub. If you place it next to his hay manger (if he has one), the bucket will be constantly full of bits of floating hay and seeds or dust. Sometimes sucking up the hay bits with the water can irritate the lining of the throat, something like drinking a cup of tea with the leaves floating on top of it—not too pleasant, I am sure you'll agree.

To keep the horses separated in their stalls, bars of metal or heavy wire gridding should be used on the top sides of the stalls.

They should be placed from the top board of the stall to the ceiling or at least high enough so that, should the horse rear, he could not reach over it with either head or feet. The top board of the stall should be about withers height.

Doors of stalls may be either the kind that swing open, as do regular doors, or they may be of the kind that roll back and forth on metal rollers. The latter is a good type for use inside the barn, since they roll back out of the way, along the side of the stall. They need no room to swing as do the other door types. Doors that open the way regular doors do are fine for stalls that lead directly to the outside, since there you usually do not

Iron bars, built directly into a door frame, are a good method of keeping the horse from putting his head out into the aisle.

need to save space. Doors, too, should be solid and securely fastened, both when opened and when closed. A door should never be allowed to swing in the breeze.

Whether the stall is empty or not, the door should never be left to swing as it will. A curious horse nosing about an empty stall may be accidentally hit and hurt by a door that is loose. The same goes for someone who is working around the barn.

Now, what about stall types? There are box stalls, tie stalls, and standing stalls. You often hear the last two used interchangeably; however, there is a difference between the two, as I will explain later on. A box stall, or a loose box as it is sometimes called, is a large, free stall in which

Heavy wire mesh is another good way to keep the horse confined, yet allow him to see what is going on in the barn.

Doors that roll on a track are excellent for use inside the barn.

the horse is untied and allowed to freely wander about, getting up and lying down at will. Most stables built today are all of loose box or box stall types. These are the most comfortable and practical for the horse.

The tie stall is a stall that is smaller than the loose box. The horse is usually tied to the manger by a rope that is fastened to either his halter or a stall collar. With a bit of practice, many horses can learn to lie down and get up in a tie stall without getting cast or becoming entangled or caught in any way. Tie stalls are usually used when a horse is dangerous to handle in the stable, such as with one that charges in his stall.

A standing stall is just what the name implies. A little wider than the horse himself (sometimes his sides actually press against the sides of the stall), the standing stall is an abomination. The horse can only stand

When open, stall doors should be hooked back, out of the way, as are these two.

where he is for as long as his master decides that he should. He cannot turn around, lie down, or move other than to shift his weight from one foot to the other. He is tied to his manger and more often than not, cannot look out because standing stalls are usually not provided with low windows. He is usually staring at a blank wall that comprises the front of his stall. Neither can he move away from his manger. Should his feed go sour, moldy, or rot he has no way of getting away from either the odors or the discomfort of inhaling them. Neither can he keep his feet dry by moving to cleaner litter.

A horse should have as much freedom as possible, both in the stable and in the field—freedom with supervision of course. Too often, horses that are housed in standing stalls develop physical problems, from swelling legs to respiratory difficulties.

Stables can be constructed in nearly as many varied designs as can houses for human habitation. There are single level, one row structures; two-story barns for feed and bedding storage overhead; bank barns that are built into hillsides for the maximum containment of body heat; barns with aisles down the center that may be used as riding areas; barns in the design of a square within a square with riding room all the way around.

The designs are endless and all have their own merits. It is mainly up to the owner to decide which type barn fits his needs best. If he has a

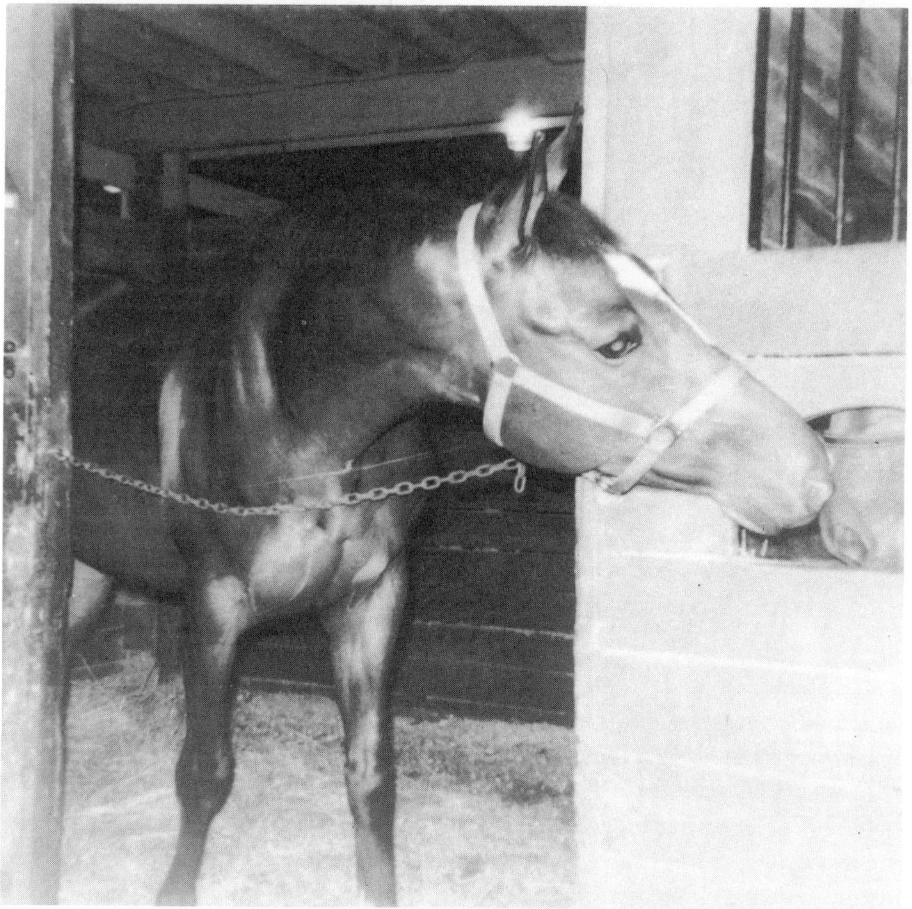

A chain, at chest height, is a good method of keeping a horse in his stall if you must leave his door open for any reason. Note the roominess of his loose box.

The barn pictured here is a one-story, three-stall barn. Each of the three stalls has direct access to a fenced-in paddock area.

one-horse stable, perhaps a two stall barn would be fine: one stall used to house the horse and the other to house tack and other equipment. If he has a large show stable, perhaps the idea of the hollow square would be good. The riding space inside would help him to keep his horses in training all year round.

Now, what about housekeeping in a barn? First of all, and I cannot stress this enough, a barn should be neat and clean. Clutter has no place around a barn. The less dangers and obstacles that horse and rider have to contend with, the better off and safer both will be. Barns that are not kept clean have other dangers besides those of tripping and falling.

Barns that have yards and yards of spider webs festooning the walls

This is a fine example of a small, two story barn where the feed, bedding, and tack are stored above.

and rafters are potential fire hazards. Webs gather a lot of dust, dirt, and after awhile, bits of hay and straw or shavings. If allowed to accumulate such debris, webs become matted, which in turn strengthens them to hold other objects, such as larger pieces of straw, hay, and more dirt. Too, webs contain a lot of dust and when they break, or if they are brushed against by the horse, this dust can easily be sucked into the lungs or drop into the animals eyes. This sets up the ideal condition for respiratory trouble and ophthalmic inflammation.

There is no place in a good stable for neglect. Cluttered aisles are another problem. Aisles in barns are for one thing and one thing only. That is for walking a horse in and out of the barn, to the wash rack or

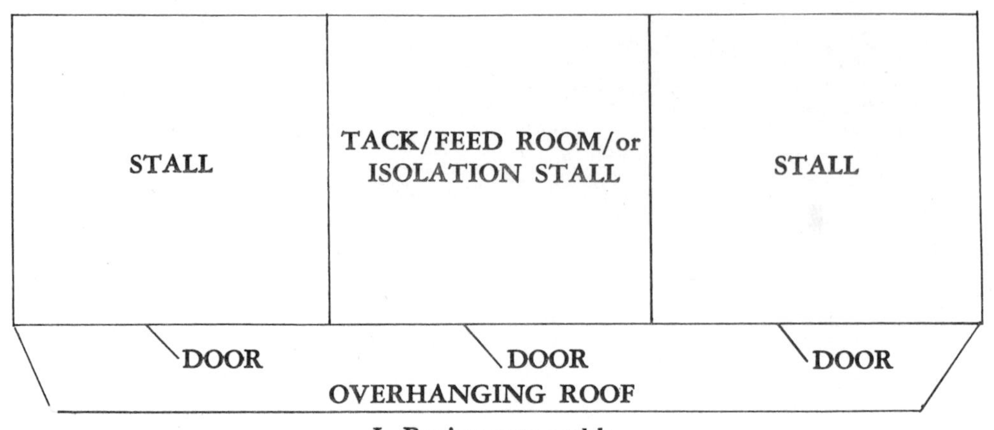

I. Racing-type stable.

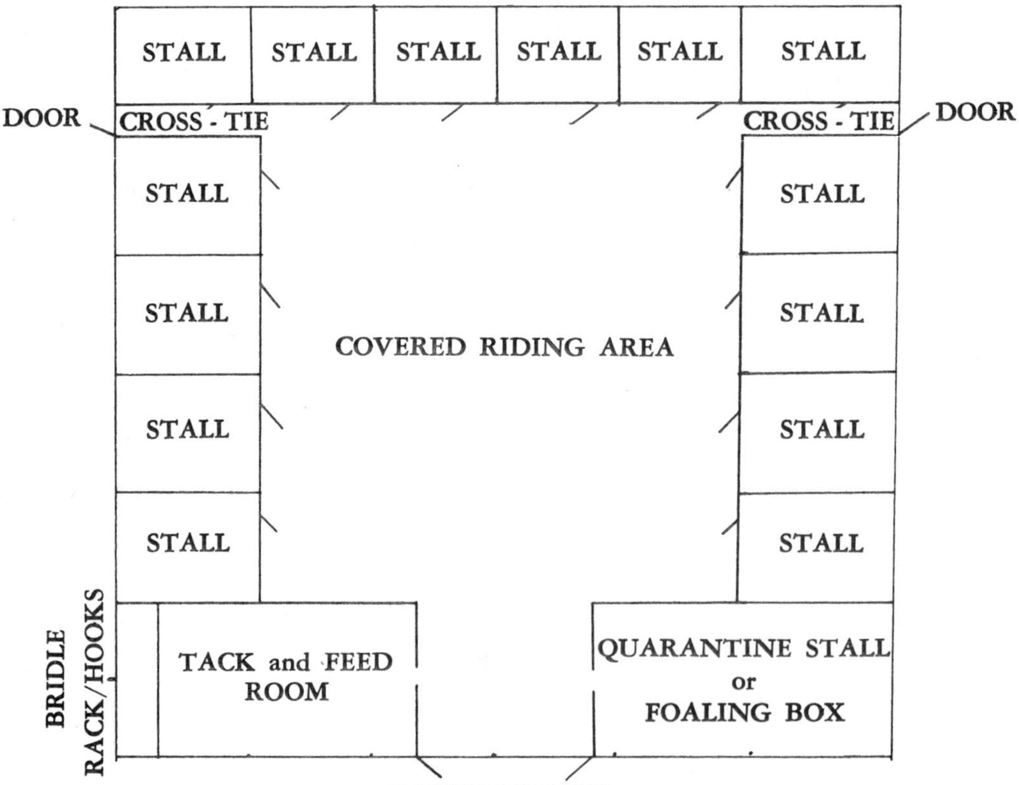

II. Racing or hunting stable.

STALL
STALL
STALL
DOUBLE DOORS
RIDING AREA
DOUBLE DOORS
STALL
STALL
STALL or TACK/FEED ROOM

III.

STALL
STALL
STALL or TACK/FEED ROOM
DOORS
RIDING AREA
DOORS

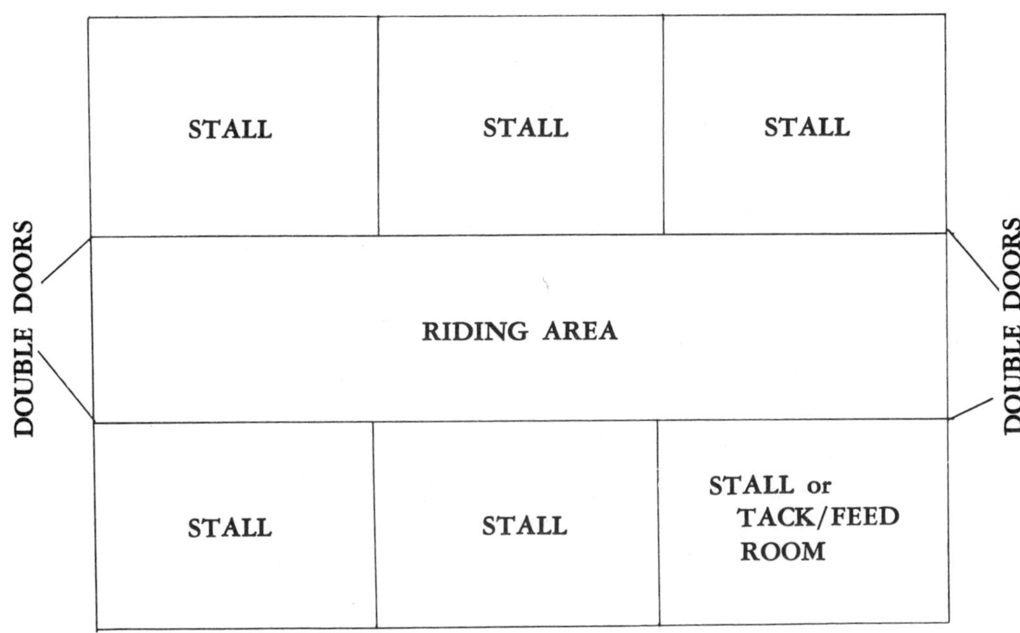

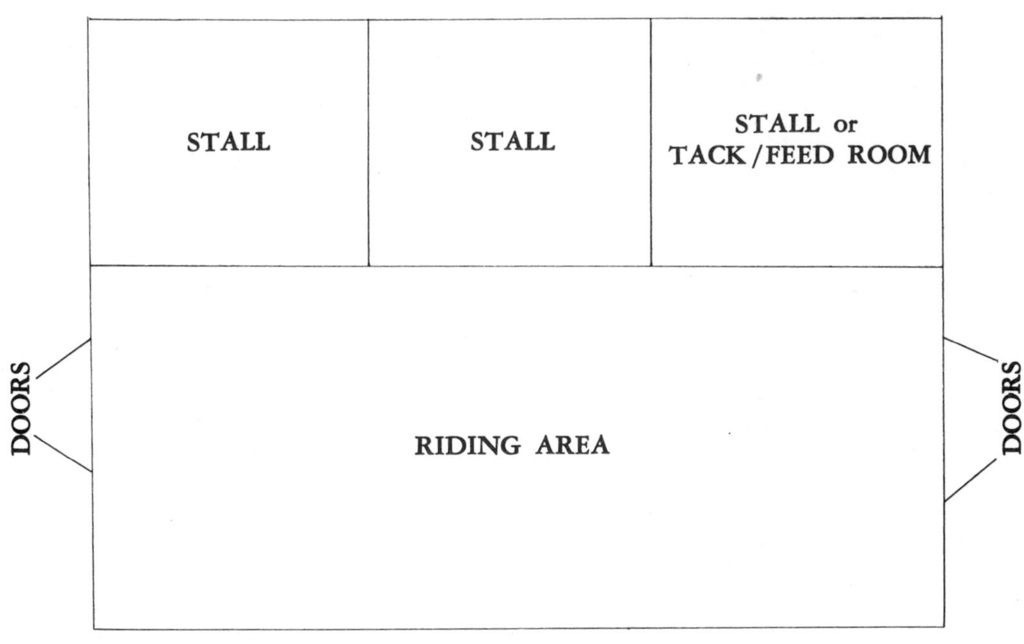

IV.

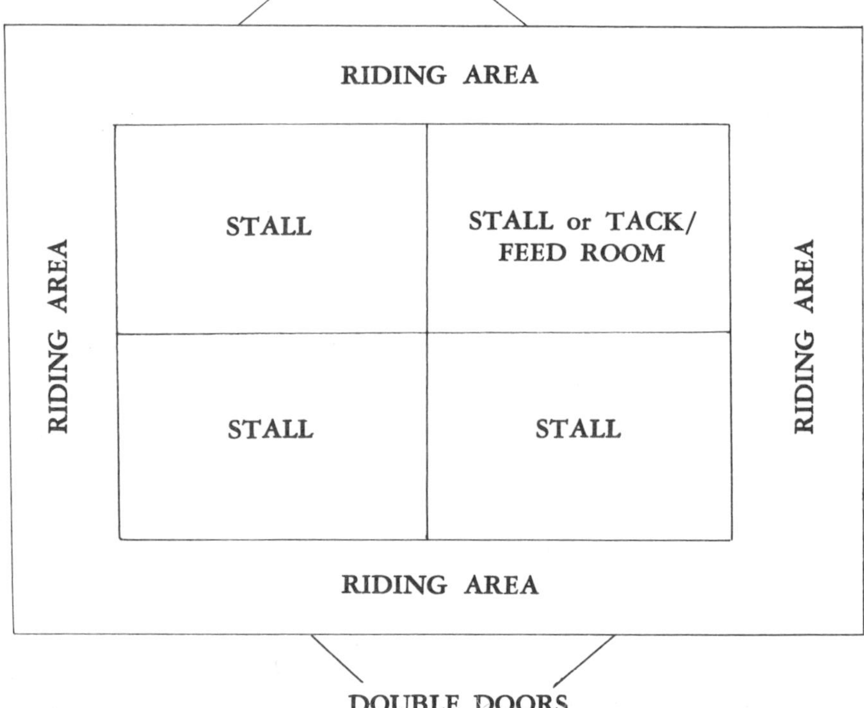

Pitchforks and other tools should never be left lying around for obvious reasons.

Heavy accumulation of webs, etc. like this is poor stable management indeed.

This stall is in dreadful need of a good cleaning and thorough disinfecting. This is an exaggeration of poor stable management.

Although this barn has a very wide center aisle, which is good, the sides of the aisle should be kept clearer.

saddling area, or to another stall or riding area. Aisles should not be used for storage of feed, hay, straw, or shavings. They should not be a repository for uncleaned tack, stable blankets, or grooming kits. They should not be littered with shovels, pitchforks, wheelbarrows, or muck buckets. The horse doesn't know that by nosing around in the stable tools he could get seriously hurt. He doesn't know that he shouldn't leave his stall on his own. He doesn't know that his door was left off the latch by accident. He can't know. You should. "Safety first" should be the rule in all stables.

Smoking should never be allowed in any barn. Large no smoking signs can be obtained nearly anywhere and they should be posted prominently, in the barns and on the outside entrances. One hot ash, dropped from a cigar or cigarette, can start a holocaust if it falls onto feed, bedding, or stable waste. Every stable should have at least one fire extinguisher. The ideal setup is to have one extinguisher in each aisle and one at the feed bin or chute, as well as one in the feed room or building used to house feed and bedding.

Every stable should be built with an extra stall, regardless of how many horses you intend to have. The extra stall should, if possible, be placed at a distance of at least one stall width from the last stall in that row. The reason for this is that the extra, single, separated stall gives you a place to isolate a horse. Should he become ill, catch a cold, or contract any diseases or ailments, he can be kept apart from the other horses,

Storing bales in aisles is one of the worst things that can be done in a stable.

thus minimizing the dangers of them coming down with whatever it is that he might have. All horses newly arrived in your stable should be kept in quarantine for at least a week; two is preferable. This gives them time to get used to your stable and it also gives them time to show symptoms of any illness or disease they might be incubating or harboring. The isolation keeps them from contaminating the rest of your horses. Too, an isolation stall makes a good foaling stall, especially if you build it larger than a regular stall. The extra size is a good idea. If a horse is injured and needs a lot of attention or needs to lie down frequently, the extra room will be welcome, both by him and by you. It is no fun treating

An aisle that is blocked by hay or straw that has been placed there, either for storage or as a disposal location, is a fire trap.

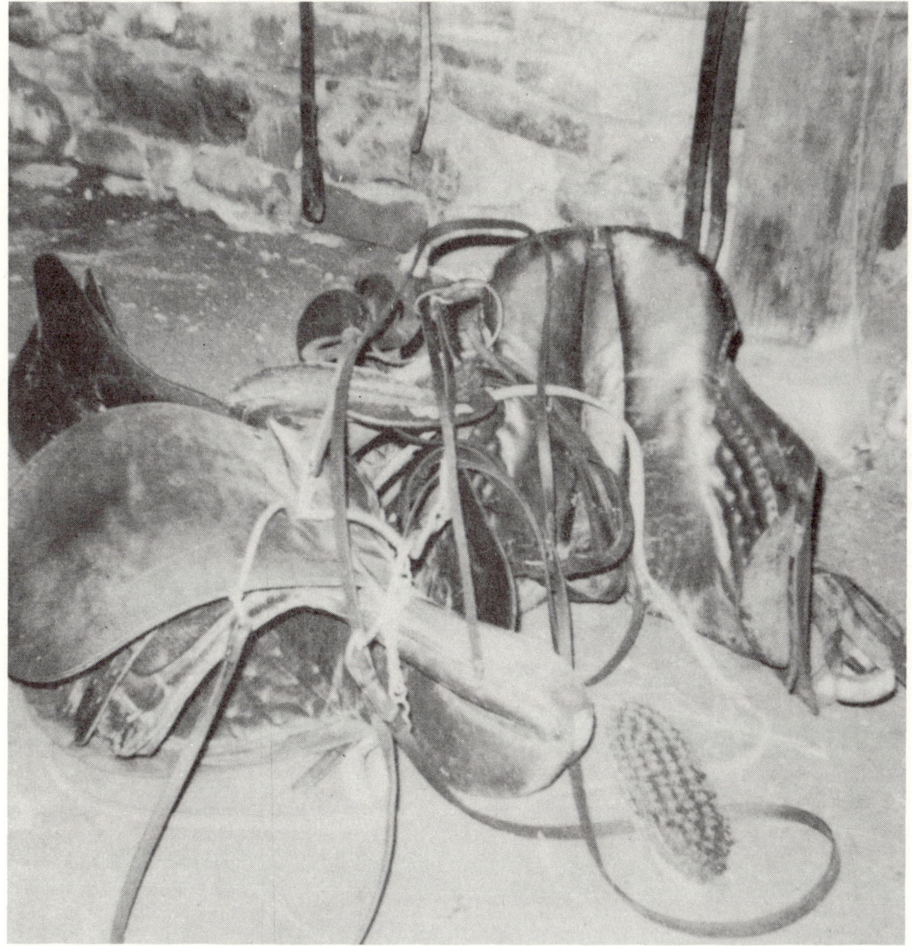

Tack should never be left in barn aisles like this.

a sick horse, especially if his stall is crowded. A set of cross-ties is also an advantage in an isolation stall. They are invaluable if a painful injury needs treatment or if the horse cannot be removed for grooming and is restive.

A foaling stall should be roomy to give the mare a chance to lie down and stay away from the stall walls. She should give birth to her foal much more easily if she is not cramped for space. Too, if removed from the other horses, even at a short distance, they are less likely to disturb her and she should be able to relax more easily when her time comes. However, move her to the new stall several days before she is due to foal so that she will have ample time to get used to her new surround-

A fine example of fire extinguisher placement. This one is located in the center of the barn aisle, on the wall.

ings. Be sure that she doesn't fret. Most mares seem to foal more easily in private; however, there are always exceptions to every rule, so when you move her, keep an eye on her. She should settle down shortly after being placed in the new stall. If she does not, and appears to be nervous or upset by the next day, it might be best to put her back into her own stall. This, however, will be the exception.

Ceiling height is another important factor in building or remodeling a barn. The ceilings of the stalls should be high enough so that if the horse rears he will not strike his head, yet not too much higher than that, since you don't want the loss of a lot of body heat during the winter. Ideally, ceilings should be between 12 to 14 feet in height.

Another factor to consider when building a stable is proper ceiling ventilation. A good thing to install is a row of louvers along the eaves of the barn. Ventilating louvers allow the heated air to pass out of the barn, thus cooling it in summer. Barn louvers are made so that they can be opened and shut, thus when closed in the winter they prevent the warmed air from passing out of them. On mild winter days you may wish to open them and allow the stale upper air to circulate out of the barn, thus preventing the problem of dead air. Dead air is air that has not been changed or circulated and it can sometimes cause a barn to become colder than the outside temperature.

An alternative to this is to have ventilating cupolas installed on your

A very well constructed wash rack and cross-tie area. It has its wooden shelf to the left, lights and a drain in the floor.

This shelf in the cross-tie area holds tools needed for grooming.

barn roof. They are especially designed to allow air to circulate out, and yet keep cold air from blowing into the barn.

Another addition to your stable is the wash rack or cross-tie area. This is usually constructed with a concrete floor, complete with center drain. It contains a pair of cross-tie chains, most often a florescent bar light on each side, and sometimes a shelf at one side to hold grooming tools. Its purpose is to give you a place to groom your horse, outside his stall, where he is controlled. It may also be used by the blacksmith when shoeing or by the veterinarian when treating a sick horse. Here, a horse may also be washed down, trimmed, clipped, or tacked up for a ride. A cross-tie area is a most useful part of the stable.

A last addition to your stable is a medicine cabinet or chest. Ideally, this should be kept in the feed room or tack room where it is out of reach of strangers, visitors, and the horses. All medicines, liniments, washes, bandages, etc. should be kept in the cabinet. There they are handy, kept clean, are out of the way, yet you have ready access to all of the things you need. No stable should be considered complete until it has had its medicine cabinet installed and all necessary preparations put into it, ready for use.

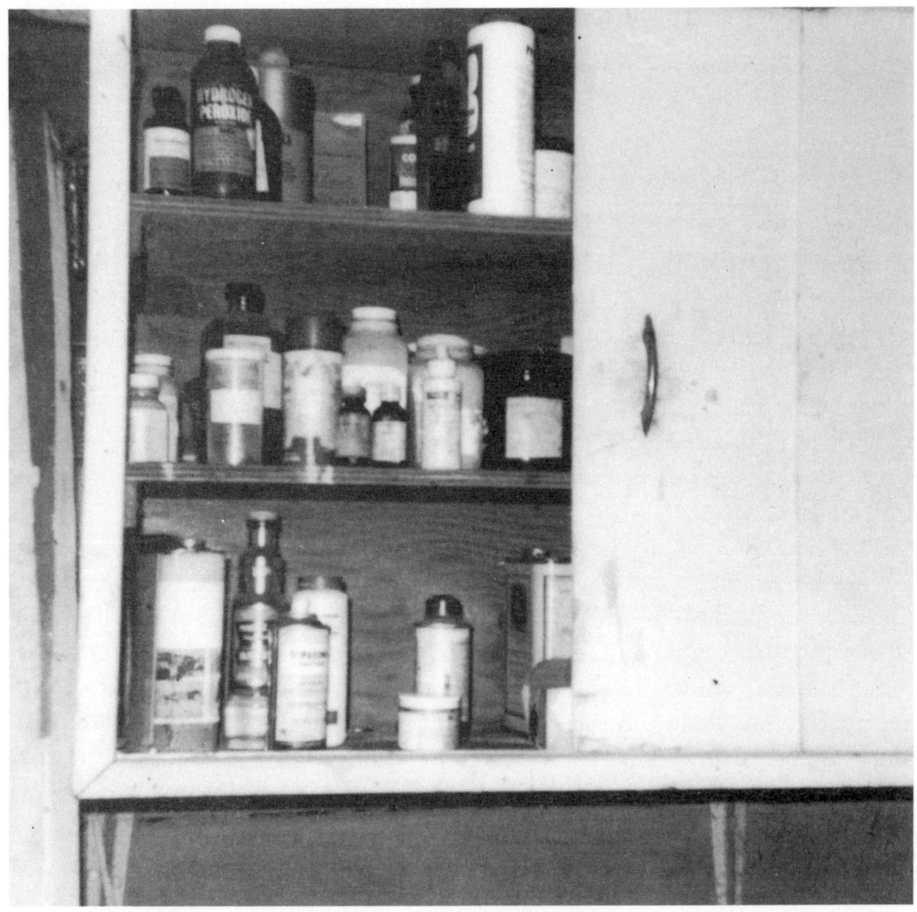

This well stocked medicine chest hangs at the end of one aisle of the barn. All medicines are labeled and ready for instant use.

2
FEED

WHILE NO ONE THING in stable management should be more important than another, feed is one thing that should be thoroughly understood. Full knowledge of just what constitutes a good feed, a nourishing feed, and just which feed is right for your horses, helps to make you a better stable manager.

Let's look at the different types of feeds and supplements and see just what should be known about each. Let's also see how to buy by quality, smell, and texture.

First let us look at oats—plain, ordinary oats. Oats are a basic feed for all horses. Because the horse in an ungulate or grass eating animal, the oat is a natural food for him, since oats belong to the grass family. When buying oats you should first look at the quality of the grain. When mature, an oat will be plump, golden, and full looking. The kernels should appear full, smooth to the touch, be shiny and slightly slippery in the hand. *Never* should you buy oats that are powdery dry, grey in color, flat kernel ed, or very dusty. Before you buy, smell the oats. They should have a nut-like aroma and have a distinctly rustly sound when shifted around in the bin or bag. Oats should never smell of mold, mildew, or dampness. Neither should they have no scent at all. You can tell if an oat is moldy by the way it feels and often by the way it looks. A moldy oat will be sticky to the touch, the kernels clinging together. The oats will not rustle together when moved. Oftentimes the kernels will be black or grey looking and will leave a residue on the hands after being held. Also beware of oats that have oat rust or blight. These oats will have a powdery red residue that will be left on the hands after being held. Any molds, mildews, or rusts can cause serious damage to your horse if fed to him. He can colic, founder, lose his sight, etc.

Another thing to avoid when buying oats are old oats. Old oats will be flat kerneled, be very dry or powdery to the touch, and while they will rustle together when shifted around, they may be grey in color. Unlike moldy oats, however, they will not leave a sticky residue on the hands. Any residue they leave will be a dry, light powder. Old oats should not be fed for the simple reason that they are sure to have lost most of their food value and nutritional properties. Oats that are stored for longer than a year are poor feed for the horse. When buying feed, ask for the current season's grain, if possible.

When buying new oats, the only thing to watch for (other than blight and mold) is that the oats are ripe and not still slightly green. Oats that are not completely ripe can cause severe colic and other complications if fed to the horse. Green or slightly green oats will lack the golden, shiny appearance of well ripened feed. The kernels will be dull, may be tinged with green, and may feel as though each kernel were slightly serrated when rubbed together in the hand.

The ideal horse oat should be plump, golden, smooth, and slippery to the touch, dry, but not parched to the touch, should leave no residue on the hand, should rustle when moved about, and should have a good, nutlike aroma.

Oats are also processed in feeds. The next type of oat is the processed one, which is also used by itself. This is the crimped or crushed oat. Many people advocate using crimped oats because they are more easily chewed by the horse and more easily digested. However any crimping or crushing of the oat bursts the hull or husk of the oat. When the husk or kernel of the oat is broken, air gets into the grain and the vitamin and mineral content begins to disappear. This is much the same as vitamins for humans. Almost all human vitamins are sealed in a sugar coating. This is not just to eliminate the bad taste of the vitamin, but to help seal the vitamins and minerals into the pill, until it is used. The kernel of the oat is nature's way of sealing the vitamins and minerals into the oat until it is eaten by the horse or other animal. Rather than having you use crimped oats because of the ease of chewing, I would suggest you have your veterinarian float your horse's teeth so that he can chew the way he is supposed to chew. Floating is like going to the dentist. The veterinarian will file off the long edges of your horses teeth, if he is cutting himself on the inside of his cheek, smooth off any rough places that may hinder his chewing, and trim down any teeth that are too long. (A horse's teeth never stop growing, so this is easy and painless to do.)

If you have your horse's teeth floated once a year, you should not have need of the crimped oats and your horse will be more comfortable besides. If you insist on using the crimped oats then by all means you should feed a vitamin supplement, too.

The next feed is mixed feed, often called *molasses oats.* This is a feed made up of a variety of things. Usually it is oats (often crimped), crushed corn, soy beans, sometimes barley, some bran, and is all held together with molasses base. The above formula may differ with different brands or different feed mills. However, most have vitamins added, as well as minerals, and most are good, all-around feeds, most of which are used the same as you would use oats. Some feed mills will mix the feed for you if you want more of something added or something else deleted. Some will have more molasses in the base than others and it will be up to you to decide which is best for your horse.

In any event, when buying mixed feed, check for the following. Smell the feed. It should smell appetizing, rich, and sweet. When held in the hand it should just be slightly damp to the touch, leaving tiny flakes of itself on the palm after being held. If the base is heavy in molasses, it may hold together in the hand after being squeezed. If the molasses is lighter in the base, it will not cling together as much. In any event, the feed should be fresh, sweet, and free of mold and mildew or other foreign objects, such as sticks and twigs, bits of paper, or twine ends.

Mixed feed, when moldy, mildewed or old, will often smell sour, like vinegar or silage. It may either be very sticky, along with the sour smell, or it may be dry and cardboardy, if very old. The color of mixed feed that has either gone bad or that is very old will range from grey to black, depending on whether it is old and mildewed or newly wet and rotting or fermenting. Under no circumstances should feed in these conditions be fed to the horse.

Sometimes in the choosing of a mixed feed for your horse you should consider what feed he has been used to and then start from there. If your horse has been on a straight oats feed and you want to switch him to mixed feed, try him on one with a light molasses base first. As with any change of feed, do it over a period of time, preferably over a period of one month. Slowly cut down the amount of oats he is getting and gradually increase the amount of mixed feed, keeping to the usual amount to which he has been rationed at each feeding. Keep track of his weight. If he starts to gain weight, and you don't want him to, cut down on the amount he is getting at each feeding. Some horses have a tendency to gain weight on mixed feed. After he has been completely switched, if you are not satisfied with his performance, the bloom of his coat or his overall condition, you may want to feed a richer, heavier mix. Again, change slowly.

Sometimes a heavy molasses mix will cause the horse to sweat heavily or, in very hot-blooded horses, may produce small pimples on his skin. If this occurs, cut out the heavy molasses mix and try a lighter base.

In any event, watch your horse carefully with any feed change. Some horses will do very well on just oats or on oats and a vitamin supplement. Others do much better on mixes, some on light mixes and some on heavy. The type and amount of work done by your horse will also be a big factor in determining which type of feed is best for him. Usually, the more and harder the work your horse has to do, the richer the feed he will be able to tolerate. As with anything, there are exceptions to the rule. Horses that have been foundered for instance, or ones that colic very easily, will ofen be unable to tolerate the heating properties of mixed feeds and will need straight oats. Also, horses that have a tendency to gain weight or to perspire easily will do better without the mix.

On the other hand, those horses which will not hold a steady weight and always need more to eat to look good will often do well on a mix. The same is true with horses that need extra oil for their hoofs and coats. Most of these types do well on a mix.

For a horse that seems to need more than oats, yet less than a mix, it might be worthwhile to try a feed mixture that is used in Great Britian. It is basically oats with barley and split dried beans or lentils mixed into them. This contains the oils needed for hoofs and coats, without the heating properties of the corn and molasses found in the mixes.

Regardless of which feed type you use, remember, the horse has a small stomach and should never be fed too much at one time. Small quantities spread over a day's time are much better than feeding one large amount once a day. The ideal way of feeding is to split the grain portion into three small meals; however, the most often used and accepted method of feeding is to split the daily ration of grain into equal morning and evening feedings. Depending on your particular horse's condition, you yourself will have to decide just how much or how little should be given at each feeding. (The best thing is to check with your vet for size of portions to be given at each feeding.) Remember though, DO NOT OVERFEED. It could have very serious and detrimental effects on your horse.

While we are on the subject of feeds, I feel that I should mention some things about corn, the for and against factors. I believe you will find that there are more factors against it than for it as a feed for horses. While corn is a grain, the same as oats, corn is high in fat where oats are high in protein. Corn is fed primarily to cattle to fatten them for market and it is not a good idea to feed corn to horses, since it does the same thing to them. Unless your horse is doing extremely hard work, such as fox hunting every day, I advocate no corn. The only other time that corn would be acceptable for use as a horse feed is if you are trying to build up an extremely emaciated horse and need to preserve its body heat. However, any time corn is fed, keep a careful eye on the amount fed and on the way the horse adjusts to it. Corn is highly fattening and a fat or overly fat horse is prone to the same troubles and ills as is an overly fat human being: foot trouble from the weight, heart trouble from the excess fat surrounding the heart, strokes from clogged blood vessels, high blood pressure, and the inability to exercise even mildly without labored breathing, excess perspiring, and rapid heart beat. There is an old horseman's adage to the effect that "it takes a lean horse to run a long race," and this is very true. You never see a fat Thoroughbred used for racing, just as you never find a fat cross-country runner in human races. It is much easier for a horse to perform when on the thin side than it is for him when he carrying around an extra hundred pounds or so. Try this to prove it to yourself. Walk up a flight of stairs without carrying anything. Then walk up the same flight of stairs carrying a western saddle or a full sack of feed or grain. You will soon see what I mean. Pity the poor horse when he is overweight and can't put his burden down.

If you do feed corn to a hard working horse or as an occasional treat, be sure it is of high quality. When buying corn you can get it two ways: unshelled, with the corn still on the cob, or shelled and bagged. Either way, be sure it is of the best quality. Cob corn should have full kernels that should be hard to the touch.. Never buy cob corn that is soft or squashy feeling. It may not be completely ripe or it may have been soaked with water wherever it was stored. The full cob should be a deep yellow in color and shiny, and their kernels should be firmly attached to the cob. There should be no matter, other than the corn silks, clinging between the rows or kernels of corn. Black or grey-black matter between the rows or kernels may be either mold or corn smut, either of which is bad for the horse. Smell the corn. It should have little odor, but what it does have should be somewhat nutlike, as were the oats.

Buying shelled corn is often tricky because most often the bags will be stitched shut and some feed mills will not open them for your inspection. If you can inspect the shelled corn, see that each kernel is white or pale yellow where it was attached to the cob. Be sure that there is no mold or corn smut attached to the kernel root.

Some people prefer to buy shelled corn, since they don't have to get rid of the cobs. However, biting the corn from the cob is good exercise for the horse's teeth and gums, and since corn shouldn't be used as a feed very often anyway, you shouldn't have that many cobs to get rid of. The few corn cobs that you might have may be stored in old feed bags, someplace dry, and when you have gathered a sack full you can take them to your feed mill and they will grind them for you. Ground corn cobs make good mulch for gardens, flower beds, etc., as well as good litter for poultry houses. Dried corn cobs also make excellent fire starters for fireplaces or camp fires. Be sure, however, to store them away from the horses. As they dry they become more flammable, and all flammable materials should be kept in another building, other than your barn.

Now, what about supplements? There are many kinds of feed supplements on the market, so let's take a look at a few of the types. Most, if

Pelleted horse feed is an increasingly popular feed for horses. As you will notice, even foal feed comes in pelleted form.

not all, are to be mixed into the feed ration at each meal time or are sprayed onto the feed at each meal time. Let's look first at the powder type. This vitamin-mineral supplement is in a fine powder and is to be mixed with the feed. This is much the same type compound found in human vitamins that are coated and made into pill form. It is, of course, in a less refined state. The only problem with this type of supplement is that if it is not mixed thoroughly with the feed, the horse either may leave it, if it is on the bottom of the feed tub, or he may inhale it and have trouble with coughing if it is on top of the feed ration. However, once mixed well into the feed as it should be, this type of supplement is readily adapted to by the horse.

Another type of supplement is in pellet form. It is a vitamin and min-

One type of feed for horses that cannot tolerate oats.

One type of compressed horse feed on the market. There are many excellent ones available. The choice is yours, and your horse's.

eral charged alfalfa compound and is accepted readily by most horses. It is a very good supplement to use when the horse has a lung problem and has difficulty tolerating his hay.

Still another type is an oil supplement that is sprayed onto the feed at meal time. There are also supplements for pregnant mares, lactating mares, stallions in service, and growing foals.

Regardless of the type of supplement you use, be sure to read the labels carefully before you buy any of them. Be sure that the supplement you choose contains things that will be as nearly a natural food to the horse as possible. Since the horse is not a carnivore, it is difficult for a horse's system to accept or digest animal fats, so check to see that the fats contained in the supplement are polyunsaturated. Also, see that your supplement does not contain dessicated (dried) liver or other dried animal parts.

Bran can be used as a supplement, especially with horses that have a tendency toward bowel impactions, constipation, or hard stools. Bran is mildly laxative and is nourishing, so it makes a good supplement for the above types of horses. It is also good for any horse, once in awhile, especially after a very hard day at work. Make your horse a hot bran mash for supper. Place three or four quarts of bran and a handful of salt into a clean metal bucket, and pour in enough boiling water to make the bran the consistency of medium-thick oatmeal. If you wish, you may also add a handful of chopped carrots or cut up, well-ripened apple. Stir the mixture well, cover with a feed sack, and allow it to cool until the whole thing is lukewarm. Then feed to the horse as you would a regular meal. Bran may also be fed dry, mixed in with the oats. It produces a soothing, mildly laxative feed in this form and is often used like this in the initial treatment of founder or laminitis.

What about pellets and chows as feeds? Some horses cannot tolerate oats, etc. because of lung troubles (heaves, etc.). These horses can be fed hay pellets in place of their hay and special horse chows to replace their oats. These products are especially designed to be dust free, to taste good, and to be nutritious. They are usually a type of compressed mixed feed that is ground fine, charged with vitamins and minerals, and compressed into chunks that are easily eaten.

Now, how about hay? How do you buy good hay? How do you tell good hay from bad? How do you separate hay into the different types and qualities? How do you tell if it is fresh, if it is green, if it is moldy, or mildewed? How do you tell if it is fit to feed a horse? How do you tell what kind of hay it is? These are all things you should know in order to become a good stable manager.

First, let's go to buy hay. In order to buy hay you should do the same as when you bought oats or other grains. Go to the feed mill yourself and choose your own feeds. This may be a bit time consuming, but it pays in the long run. When buying hay, one of the first things to do is to smell it. Hay that is fresh will smell just that way—like fresh-cut grass, only dry. It should smell clean, sweet, and good. Hay that is fresh will feel springy to the touch, not be dry or crumbly and powder in the hand.

Since nearly all hay is baled by the time it gets to the feed mills, it is sometimes difficult to tell how old the bales are. However, your senses of smell and touch should be your best guides as to whether or not to buy the hay. Hay that feels damp, wet, or slightly rubbery may not be completely cured and properly dried. Just as with oats and grains, steer clear of green hay and other baled grasses. Green hay may ferment, mold, or mildew in the bale, and can cause serious trouble if fed to the horse.

Feel the bales well. The hay should be dry, yet pliable. The hay should smell fresh, but not green. *It should not smell moldy*. Moldy hay smells quite strongly of mold. Since the bale usually molds from the inside out and since the mold odor has nowhere to go, the smell is trapped inside and is quite strong, especially as you get close to the middle of the bale. If you can, work your hands down into the bale, as far as you can, and then smell the bale. The separation of the grasses by your hands allows the mold odor to rise to the top of the bale and escape, thus making it much easier for you to detect the trouble with the bale. It is not necessary to do this with every bale you order; however, you should

Bales of hay and straw should never be stored outside, unprotected like this.

specify to the feed mill that you don't want any hay that has been stored outside, near open loft windows, or in hay stacks before baling.

Another thing is never to buy old hay or hay that has been stored for more than one year. Old hay will be grey in color, will be very brittle, will have very little hay odor, and will usually leave a powdery residue, much like dust, on the hands when handled. Old hay, like old oats, has little nutritional value. As the grass stalks age the vitamin-mineral content is lost, little by little. Unlike grass that is left growing in the fields to be winter cured, old hay does not retain its food and calorie value. (See chapter on pastures for winter range.)

Putting the next meal's ration of feed in the containers for distribution, and then storing them in fiber drums, can be a time saver at meal time.

A large, wooden feed box, with the lid open.

Now, what is the best hay to feed a horse? As far as location, some of the best hay in the United States is grown in upper New York State, Vermont, and Montana. These locations seem to have the ideal climatic conditions and soils for producing superb hays. As far as the type of hays, the most acceptable hay to feed is timothy. Timothy hay seems to agree digestively with more horses than any other hay or hay mix. Some people prefer to feed a timothy-alfalfa mix; however, since alfalfa is inclined to be a laxative, it needs close supervision if fed regularly. Unless there is a severe case of bowl impactions or chronic bowl impaction problem, it is best not to feed alfalfa straight.

Timothy is a meduim-density hay, with long, straight stems and very small, thin heads of flowers. Unlike clover or alfalfa hay, there will be no large blossoms. The leaves of timothy are thin and straight, unlike clover or alfalfa, which has oval, short, and broad leaves. Timothy tends to be slightly less green in color when baled than does the alfalfa. If you want to feed a timothy-alfalfa mix, ask your feed mill for one that is less than half and half. That is to say, ask for bales that are at least three-quarter timothy and one-quarter alfalfa. Again, be sure to watch the horses stools to see that they are not becoming green in color or runny. Alfalfa is not well tolerated by some horses and may cause diarrhea.

Now, what is a good way to store grains and hay? Let's begin with storing grain feeds. There are many ways to store feed or grains, but however you store them, be sure that they are well out of the horse's reach or stored some way that the horse can't get into the feed containers. Unfortunately, the horse isn't very smart when it comes to eating. He doesn't realize that if he eats as much as he wants at one time he may very well kill himself. A horse may very well eat himself to death or eat enough to founder himself, if given half a chance. Put all feeds under lock and key if possible.

There are a number of good ways to store grains. The most common way is to have a feed box. Feed boxes come in various sizes and most are made of aged wood. They have a slant top that fits flush with the sides and front, thus making it nearly impossible for a horse to lift, should he get into the feed room area. The tops of feed boxes must be lifted straight up and tilted back over the box itself in order to keep them open. Thus, as you can see, it is very difficult for a horse to accomplish such a feat. Another way of storing feed or grain is to keep it in the largest galvanized garbage cans available. The metal cans keeps the feed dry and cannot be chewed through by vermin, as can the wooden feed boxes or plastic cans. The only thing with this system is to be sure the cans are placed well away and out of reach of the horse. Some horses are very adept with their teeth and may be able to remove the can lids, *regardless* of how tightly they fit. Another way of storing grain is to procure a metal overseas trunk at your local Army-Navy store and keep the feed in that. The trunk should be put up on bricks or cement blocks to allow for ventilation and to keep it off the floor, away from the dampness. These trunks can be locked and provide nearly horseproof methods of storage.

If you use a lot of feed in a short time, for example if you have a large stable, you can store your grains in fiber barrels or drums. These should be placed either on bricks or boards to keep them off the floor, to keep the bottoms dry and the feed as well. They should also have tight-fitting lids. This is not a good method for long-term storage, since the drums have a tendency to absorb moisture and this may cause the grain to become damp, moldy, or mildewed, if kept in them for long periods of

Note the slant of the top of the box. Also the chute from the overhead delivery container. The box may be filled from above with no dust being released in the barn.

time. New wooden barrels are also not a good method of grain storage for the same reason. New wood, also, if damp for a long time, may cause the grains to mold or to react with the resins in the wood, and the grain may ferment and/or rot. Grain kept in new wooden barrels may also sour and become silage like.

The best methods of grain storage are feed bins, galvanized garbage cans, and overseas chests (metal ones).

Acceptable methods include fiber barrels or drums for short term storage. Plastic may be used if your vermin population is kept low. Rats and mice like to chew through plastic.

Fiber barrels, such as these, are good for short-term storage of grains.

Unacceptable methods include new wooden barrels or containers.

Now, what about hay storage? The most common method of storing hay is to stack the bales in a place where the horse can't get at them. Again, there is a right way and a wrong way to do this. Bales should be stacked in a pattern that allows for ventilation, thus keeping all the bales dry and mold free. For example, place the bottom row of bales side by side in a double row, that is, two rows of eight bales each. Place the next two rows side by side, on top of the first, in the opposite direction as the bottom, skipping the two center bales in the pile. The third row should be placed in the same direction as the bottom row, only leave out one bale next to each end. This open method of stacking allows free movement of air all around the bales and also cuts down on the chance of fire by spontaneous combustion. Spontaneous combustion is caused when some of the bales are slightly damp. Damp hay produces its own heat. When this heat is trapped in the middle of a stack of bales and cannot get out, the temperature of the stack increases until it reaches the ignition point and a fire results. This is why I strongly advocate that all feeds and beddings be kept in a different building than the one that stables your horses. If you must keep feeds in the same building, then try to place them on the same floor level as the horses and not overhead. Hay and grain fires are particularly hot and fast moving, so for the sake of your horses, try to

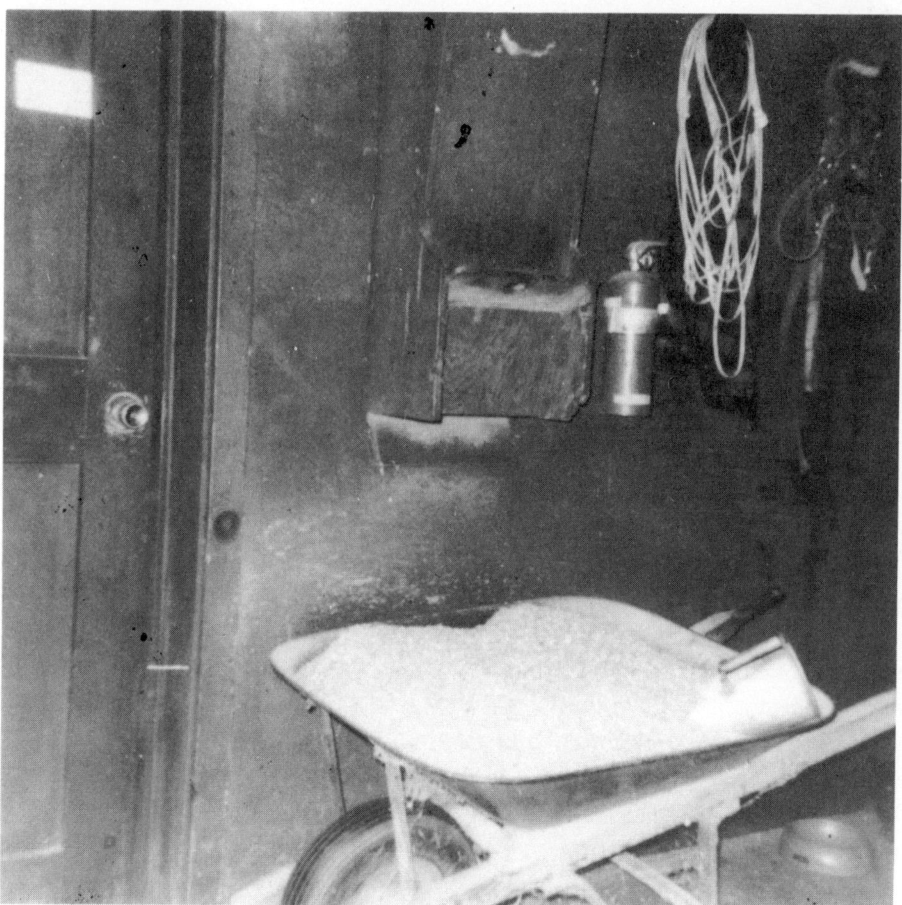

This photo shows another method of feed storage and distribution. The feed is stored in the loft. At feeding time the wheelbarrow is placed underneath the feed chute, the chute is opened, and the proper amount of feed falls into the barrow. It is a good idea to keep a lock on this type of feed chute.

These bales are properly elevated above the ground on raised planks.

be as safety conscious as possible. A fire on the same level as the stable may possibly be contained, at least until the horses are removed from the building. A fire on a second floor of a stable is nearly always uncontrollable.

Again, as with grains, see that the bales of hay are kept off the floor, either by placing them on boards or building a platform to hold them. See that lofts are weather tight so that rain and snow can't get in and onto the bales, thus wetting them and producing ideal mold conditions. In summer, keep your hay storage doors open during the day if it isn't raining, in order to produce as dry an atmosphere for the hay as possible. Close them up at night to keep out dew. Keep your doors closed, as well, if it is foggy.

Dry atmosphere and good ventilation keep the hay better and the risk of fire much lower.

3
BEDDING

IN ORDER TO BE A successful or well informed stable manager, it is necessary to know that there are several separate types of bedding that are suitable for horses. You also need to know that there are beddings that are not fit or suitable for use around horses, and the reasons why. You should know how to buy bedding, how to select it, how to tell if it is fit for use, how to store it, and how to use it most effectively.

To begin, let us discuss sawdust. Sawdust is not a very good thing to use for bedding down horses for the simple reason that it is too flyaway. Being as light as it is, sawdust is prone to getting into a horse's eyes, for as he walks or moves around in his stall he stirs up the bedding. The sawdust is easily airborne because of its light weight, and it can easily settle in the animal's eyes. This is doubly true if your horse lies down to sleep. (Not all horses do this.) If the horse lies down and switches his tail, the sawdust may be stirred up to such an extent that not only does he get the bedding in his eyes, but he might inhale it as well. This inhalation of the sawdust will naturally set up an irritation in the lungs that may lead to breathing disorders later on. If the sawdust cannot be expelled through coughing on the part of the horse (and often it can't be because it will be damp from the animals breath), chronic lung infections may be produced.

Aside from the physical problems that may arise through the use of sawdust, there are drainage problems to consider as well. Though sawdust is extremely absorbent, it is also prone to packing or caking. As the sawdust gets wet from the animal's urine, it settles into itself. As the horse walks around, his weight adds pressure to the wet bedding and com-

67

presses it. This compression of the bedding does not allow the urine to penetrate to the dirt floor of the stall below and be absorbed and then drained away. The wet matter is held in the sawdust instead. This retention of the urine in the bedding permits a great deal of the ammonia vapors of the urine to collect and rise, and this, too, may cause trouble. Often large amounts of ammonia vapors will irritate a horse's eyes, to say nothing of the way it smells, to him and to you. The packing of the sawdust also poses a problem when it comes time to muck out the stall. The sawdust can hold an amazing amount of moisture and can become unbelievably heavy. Because of the weight of the sawdust when wet, it is often necessary to take less out at a time and to make more trips than normal to the disposal area. All of the above should be reason enough to discourage you or anyone from using sawdust as bedding. The only time I think that it is permissable to use sawdust as bedding is if it is well mixed with wood shavings, and then only if there is no substitute bedding available.

Since I have mentioned it, let's talk about wood shavings next. Wood shavings are, to my way of thinking, the best possible horse bedding. These are shavings of wood from planing mills and lumber yards that do their own mill work. Cabinetmakers and carpenters, too, often have excess shavings of wood to get rid of or dispose. The shavings are really large flakes of wood, thin, often curled from being planed from a larger piece of wood. By large flakes I mean the size of a quarter to the size of a silver dollar. Wood shavings are absorbent, yet do not pack or cake like sawdust. They permit all liquid matter to soak through them, retaining very little themselves, to be absorbed by the stall floor and drained away. Shavings make it easy to see the stools dropped by the horse and this allows quick disposal of them, since they can be removed from the stall whenever they are noticed. Droppings should be removed at least once a day. If the horse is stabled over night, then his bed should be freshened when he is out to pasture during the day. It will then be ready and clean for him when he is put back into the stall at night. If he is put out to pasture at night, such as in the summer, then the droppings should be removed as soon as he is put out. It will then be ready when he comes in for breakfast in the morning.

Shavings are light in weight and because of their draining the liquids away, except for normal absorption, they remain fairly easy to handle when cleaning out the stall. Unless the horse remains in his stall all the time and unless the stall is not cleaned out for a long time (and this is definitely not good), there is little compression or packing of the bedding. If your horse is very active in his stall the bedding may become ground up quite a bit, but it still will not pack or cake the way sawdust does. Unless they are old, shavings do not hold nearly the amount of ammonia that sawdust does either.

Shavings often may be had for the asking from planing mills or cabinetmakers. Many are glad to have a place to dispose of their shavings. Others will charge a very nominal fee for them. Some places will charge if they have to put the shavings in bags, but will allow you to take them free if you bag them yourself. Shavings may also be purchased in bales. These are usually very high-grade shavings, baled in heavy paper and compressed, like the compressing of trash in trash mashers. The price will

These bales of stall litter, though not entirely of wood shavings, are one method of buying stable bedding. It shows the method of packaging of stable litter.

vary on these baled shavings, depending on your locale. It is, however, sometimes more economical to get your own shavings and do the little extra work of bagging them yourself. It can get to be a bit expensive buying baled shavings, especially if you have more than one horse to care for and keep clean.

Shavings also keep the horse cleaner than sawdust does. Sawdust has a tendency to get in the horse's coat and make him look dusty all the time, which he will be. Shavings, on the other hand, are easily shaken free from the horse's coat by his natural movements and they will not cling to him with the tenacity of the sawdust.

Now, how do you tell good bedding (sawdust-shavings mix or straight shavings) quality from bad? One of the best ways is to smell. The other is to touch. Sawdust, though not a good bedding, is used by some people. If you must use it be sure that you are getting the best. All bedding made from wood products (sawdust and shavings) should, first of all, never smell sour. Wood has a natural odor of its own, somewhat like green leaves, only stronger. This is fine, but do not buy shavings that smell like vinegar or ones that smell as though they were decaying. All wood will rot and, naturally, the finer the pieces of wood and the thinner it is, the quicker the rotting process will take place. Shavings that have been left outside or in wet places may begin to decay before they are used. Be sure that the shavings you get have not been piled outside for any length of time or that they have not been wet or rained upon.

Another way to tell if shavings are alright is by touch. Shavings that have been aged indoors will be dry and fluffy to the touch. They will have little or no odor and they should rustle freely when moved. New shavings, on the other hand (shavings that have been freshly milled), will have the green leaf smell and may be slight damp to the touch. They will not roll about quite as freely as the aged shavings. New shavings are like this because they still contain a minute bit of sap from the tree from which they were planed; however, if you store them properly when you get them home, they are just as good for bedding as aged ones. Just be sure to smell them. New shavings and decaying or sour ones smell quite different. If they do not smell offensive to you they will be alright for bedding use. If you bag your own shavings from a planing mill you may find that sometimes you will get cedar shavings. If this is the case, be sure to mix these shavings with others before bedding the horse down with them. Some very thin skinned horses may develop a skin rash due to the acid quality of the straight cedar. This is not to say that cedar does not make good bedding. It is just as good as any other wood, if mixed with other shavings first, before being used. Cedar shavings certainly make a stall smell good, like the inside of a cedar chest.

Shavings being stored in an empty stall. Note the fiber barrel that is used to carry the shavings from the storage place to the stall where they will be used. It is large, but light in weight.

Now, how do you store shavings and keep them dry? A dry stall that is not in use is an ideal way to keep shavings. Simply pile the shavings in the stall, close the door, and leave them until you are ready to use them. It is good, however, to check on them now and then to see that they are not getting damp. Also, if they were newly milled shavings, it is good to turn the pile every now and then. This allows the bottom of the pile to stay dry.

The best kind of stall for storing shavings like this is one with a closed bottom half and a barred upper half. The closed wooden bottom acts as a bumper and keeps the savings from spilling out onto the floor, while the open or barred upper half allows plenty of air to circulate around and into the pile.

Another way to store shavings, but for short-term storage only, is to keep them in fiber drums. These drums usually hold enough shavings to fill one third of a stall to the proper depth for bedding down. This means three trips for shavings and you are done bedding down the stall. A drum full of shavings is not too heavy for a woman to handle, or even for a large child, when making up a horse's bed.

If shavings are kept this way, the top of the drum should be kept on tightly if the shavings are seasoned and dry. This keeps the moisture out. If the shavings are newly milled and damp, however, it is best to leave the top off the drum so that they have a chance to lose some of their

Fine example of well baled straw.

moisture through evaporation. Dampness in shavings is the first step toward having them sour or decay. Shavings should not be kept in these drums for more than a week without being turned to allow for ventilation, and the drums should be up off the floor to keep them as dry as possible.

Regardless of how you store your shavings, and you probably have your own favorite method, if you use them for bedding, be sure they have plenty of ventilation and are kept dry.

Another, and very commonly used bedding is straw. One of the most commonly used straws for bedding is wheat straw. This is the dried stems of the wheat plant that are cut and baled after the wheat heads have been removed. Wheat straw is pale yellow to golden in color and shiny and the best wheat straw will readily flake out from the bale when put down for bedding. As with any bedding or feed, you should select the best. Straw that is grey, straw that clings in mats in the bale and will not flake onto the floor when shaken out is poor bedding indeed. Straw that has been wet or that is moldy will often have this clinging tendency. The pats, or sections, of straw appear to cling or stick to the rest of the bale when you try to remove them to bed down the stall. Often the inside of the bale will be dark, damp, grey, and very moldy smelling. If the mold has dried, which it may have done if the bale is old, the spores of the mold will fly about when the bale is pulled apart or when the pat of straw is shaken onto the floor of the stall. Never buy moldy straw, just as you should not buy moldy hay. Some horses eat their bedding, and eating moldy straw can cause as many problems as eating moldy hay. Wheat straw makes the best straw for bedding, since this is the least appetizing to the horse. It has little flavor and very little scent to tempt the horse.

All straw should flake apart easily when the pats are shaken. That is to say they should not cling tenaciously together when you try to shake them out for use. Wheat straw, as mentioned before, will be shiny, golden, slightly slippery, and very rustly when shifted about. It should not be excessively dusty or dry and powdery. It should never be wet or moldy.

If your horse does not eat his bedding, though many horses will do this, you might try oat straw as bedding. Oat straw is softer than wheat straw and will not be as shiny or golden in color, sometimes even being on the brownish side. The stems of oat straw are thinner than wheat straw and the straw will not rustle as much as the wheat straw when flaked from the bale. When walked on, wheat straw gives off a sound much like that of fallen leaves. Oat straw sounds much softer.

Rye straw is the least popular straw, at least as far as durability is concerned. Oat straw is not too popular because many horses will eat oat straw when they would not eat the other straws. Rye straw is the hardest and often too brittle to use as a bedding. It is yellow in color, but has little shine. It is hard to the touch and quite stiff and twiggy feeling with little absorption qualities. Because of the very tubular construction of the stems, when walked on it is easily ground away by the horses feet. The stems are nearly rigid, with little flexibility to them.

The least acceptable bedding is cut, baled field grass. Some people think that cutting their own field grass and using it for bedding will save them money on bedding. While this is true, it will be costing them more in time and effort to keep their horse's stall, and possibly their horse, clean. Field grass is extremely soft, as beddings go, and needs daily changing of the complete stall, right down to the ground. Too, most field grass is

Baled straw may be stacked like this for short periods of time, as long as they are to be used within a week. Note the open loft door, for ventilation.

fairly dark when dried and mature, and manure is not easily found in it. Also, it packs down very easily, the fine stems becoming imbedded in the manure, and it is a very wearisome burden to remove daily from the stall. It also traps the moisture from the urine and it is heavy as well as messy and odiferous.

Never use ground corncobs, mentioned in the feed chapter, for bedding for horses. Many horses cannot tolerate the dust that is associated with them. Also it has a tendency to get in their eyes and up their noses, especially if the horse lies down to sleep. Also, if the cobs are not ground well enough or into small enough pieces, lumps of cob may become lodged in the cleft of the frogs of the hoof, causing lameness if not found quickly.

Regardless of which bedding you use, though the preferred ones are wood shavings and straw, buy only the highest quality available.

Now, how do you store hay? The best method, naturally, is to store it in a place that will be dry and yet well ventilated. The bales should be stacked so that they are crisscrossed in rows with bales left out here and there to allow for circulation of air. For a good method of stacking, see the feed chapter and the piling of hay bales for storage. Straw may be stacked exactly the same way.

As with feeds, straw or bedding of any type should not be kept above the horses on a second floor level of the barn. If at all possible, try to store all flammable materials such as feed and bedding in a separate building, in order to cut down the fire hazard. Also, when piling hay or straw, place the bottom bales on planks that have been raised a few inches above the floor. This keeps the bottom bales dry yet ventilated so that they can be used as safely as the top bales without fear of mold or damp ness. This way there are no bales wasted.

4

CLEANING THE STALL

BELIEVE IT OR NOT, there is a right way and a wrong way to clean stalls. Waiting around until the manure collects to a depth of a foot or better, and then cleaning it out, is poor practice, though it is done more often than one might think. Putting fresh straw on top of old droppings, thus burying them from sight, is also no way to keep a stall.

There are no two ways about it: manure is manure—a biological, bacterial, waste product. Some people advocate leaving manure in the stall during the winter and adding straw to it now and then, just to keep the horse clean, because of the heating properties of the manure. True, manure does become hot, but only because it is going rotten. While it does give off heat, it also is one of the prime causes of thrush or hoof rot. Standing in hot, wet, bacterial laden manure is one of the worst things for a horse, except in extreme conditions of hoof dryness.

A horse is not going to die of the cold, unless he is allowed to remain out in a week-long blizzard or allowed to stand untended in a closed up barn for a week. Nature provides the horse with the proper length and thickness of hair, at the proper time of year. If he is unduly cold, nature will provide him with more coat, until the extreme is compensated for. His coat sheds in the heat doesn't it? It is logical then that it will also grow to the proper length and thickness required for winter survival. If you feel that your horse is too cold, you might buy him a stable blanket, many of which are heavily lined. However, sometimes a blanket will keep a horse too warm and he may shed prematurely. If this happens and you discontinue the blanket, he may be prone to catching cold. If you don't want to blanket, and don't mind the fuzzy look your horse gets in

Notice how untidy and unkept this stall appears. This is from putting new straw on top of old droppings.

winter, don't start. If you want to blanket and have a fairly short-coated horse for the winter, then don't discontinue blanketing in the middle of winter. Heavy manure is not going to make that big a difference in whether your horse survives or not.

Now, just how does one go about properly mucking out (manuring) a stall? Well, to begin with you should remove the horse from the stall. Put him out to pasture or in the paddock. You cannot clean a stall out to the bottom properly with a horse tramping about in it. He will be in your way and you will be in his. One of you will most likely get stepped on, poked, or knocked down (probably you). Now, first of all, see how much dry bedding there is in the stall. Dry, unsoiled bedding should be piled in one corner of the stall for later use. This is true whether you are using shavings or straw for bedding. Thrift is another thing to learn in the stable. If you can use it, don't waste it just because you may have more.

Now, proceed to rake, shovel, or pitchfork all of the soiled bedding and droppings into either your wheelbarrow or a muck basket. Muck baskets or muck sacks are seen in Europe and Great Britain more often than here in the United States. A muck basket is made of hand-woven wooden splints, usually oak because of the durability. Most have rope handles and hold about two bushels. They are very convenient if you have a lack of space or if the terrain of your property makes it difficult to push a wheelbarrow. Too, it is up to the individual preference. If you would rather carry than push, then the muck or manure basket is for you. A muck sack is easily made by splitting open a large, heavy-duty burlap bag. You simply load the manure in the middle of the open sack, fold the corners together, pick it up and haul it off to the manure pile. (The terms *mucking out* and *manuring out* are interchangeable, the difference being the origin of the terms. *Manuring out* is an American term, while *mucking out* is most often heard or used in Great Britain.)

Now, after you have made as many trips as necessary to get the stall cleaned out, you should be down to the dirt floor. Move the pile of clean bedding out of the corner and clean out underneath it too, if you did not do so at the start. Remove the clean bedding and stand it in the aisle or in the now empty wheelbarrow. See that the stall floor is flat, with none of the rocks from the pit underneath protruding. Be sure that you have cleaned the stall corners well. You should now lime the floor well. By spreading sweet lime on the stall floor you cut down on the odors considerably. It keeps the ground floor sweet and fresh and helps keep the odors from the absorbed urine down to a minimum. Spread the lime well, into the corners too. Spread it evenly, but not too thickly. You should clearly be able to see the lime on the floor, but it should not be so thick that it can be forked or shoveled off.

Next, take the bedding that you have saved and place it in the middle of the stall and start the bedding down process with it. Be sure that it is all dry and unsoiled. Now add as much new bedding as needed to complete the procedure. Sweep your aisle, put your tools away, and you are all set to install the horse in his newly cleaned stall.

By mucking out in this manner, you become efficient, the work becomes routine, and you will find yourself doing the work in less time and in a better fashion.

Too many people allow the mucking job to go undone for too long a time, simply because they cannot seem to get organized. Trying to muck

Very poor management. Manure and old bedding is literally overflowing from this stall.

out an entire stall and carry the manure out on a wheelbarrow in one trip, to save time, is ridiculous, although it has been tried. Making two, three, or four trips to the manure pile is better than breaking your back and your willingness to do the job with one herculean effort. In stable management there are no short cuts. However, learning to do a thing right can make the work go a lot faster and easier.

Now, what about the daily cleaning out? This is one of the best things you can do, both to save on bedding waste and to lessen the size of the task you have at the end of each week. Each morning, when you go to feed, check the horse's stall for droppings that he will have done overnight. Keep an old bucket handy for this. There's no sense pushing the

Some of the tools used for stall cleaning.

old wheelbarrow to the muck heap if you don't have to. As you find the droppings, pitchfork them into the bucket, or into the manure basket, if you'd rather. Be sure to shake off as much clean straw or shavings as possible.

When you've got them all, fork the bedding around to freshen it for the day. When you feed at night you can use the same procedure, also freshening the bedding. If your horse is in pasture during the day, then you only need to do this once a day, but do it as soon as you can. Manure that stays in bedding all day will often foul or contaminate other bedding, bedding that might have ben used if the manure had been removed at the beginning of the day. If your horse is in pasture at night, simply reverse the process.

Once a week the whole barn should be gone over, windows cleaned, the

Two types of pitchforks that may be used in stall cleaning.

webs removed, and things generally straightened up. Just like housekeeping, horsekeeping needs routine work.

Tools needed for stall cleaning are as follows:

One pointed-nosed shovel
One blunt-nosed shovel
One rake
One broom
One or more pitchforks
Wheelbarrow
Muck basket or muck sack

A fine rake and a broom are also necessary for good stall cleaning.

VI. Manure basket.

5

PEST CONTROL

THERE ARE MANY METHODS of pest control on the market today and many of them are produced just for the stable and barn areas. By pest control I mean all type of pests, from insects and parasites to rodents.

Insect control is one of the most necessary and vital things you can do around your stable. Flies, mosquitoes, gnats, and other insects can make life miserable for your horse if left unchecked and uncontrolled. Flies have a tendency to pester the horse's face, especially the eye and ear areas. Others like the nostrils and legs. Mosquitoes not only bite the horse for his blood, but transmit diseases, some of which are fatal to the horse.

Rats and mice also carry disease and can deplete grain and feed supplies at an astounding rate of speed.

Spraying is one of the most effective methods of ridding the stable of insect pests. Spraying for stables can be purchased at most farm supply stores and tack shops. The directions should be read carefully and the contents used accordingly in order to use them safely and for maximum effectiveness.

Fogging of the barn is another method of spraying and one of the most effective insect removal methods available. The fogger is a machine, something like a paint sprayer, that holds a liquid mixture of insecticide. It sprays a fine mist, or fog, of the insecticide throughout the stable. It allows the mixture to penetrate and thoroughly destroy all insect life in its range of spray. Of course, all livestock should be removed from the barn during fogging procedures and should not be restabled until the air in the barn can be safely breathed, with comfort, by humans (usually about twenty minutes after fogging has ended). Feed, too, should either be removed or heavily covered to prevent a residue of the spray from contaminating it. Water, too, should be removed from the stalls and not be replaced until the fog has completely settled. This is one reason to use shavings for bedding. The horse is not likely to eat shavings and pick up spray residue, as he might do with straw bedding.

Another method of control of insects, especially between foggings, is to use a sugar bait for the flies. This is a granular type bait that is set out in small dishes, out of reach of horses and children. It attracts the flies, then poisons them after they have eaten it. As with any insect killer, all dead insects should be swept up and destroyed as they are found, though they will not often be eaten by the horses, as many people believe. However, their dead bodies will attract other insects and you are trying to get rid of as many insect pests as possible, not bring more.

Odor is another pest of the barn. Stable odors are really not that hard to lower, if things are done properly. Stable cleanliness is one of the best ways to keep odors down. If you have a clean stable, one that is taken care of every day, there will be little disagreeable odor to worry about. Feeds that are kept dry will not smell moldy or musty, simply bcause they are not. Sour or wet feed is sometimes worse smelling than old manure. Stalls that are sweet limed each week at total clean-out time will not smell nearly as strongly of ammonia as will those that are just mucked out and not limed. Feeds left in feed tubs to sour are other odor causers, as are dirty feed tubs. Weekly cleaning out of feed tubs and a good scrubbing help to eliminate odors here. Also, there is a barn sweetener available on the market that may be sprinkled around the stable to help reduce odors. Bad odors help to attract insects, too, and you don't need that.

Worms, though not a pest of sight or smell, can be a real menace to the horse. Various types of worms can infest your horse, no matter how good a stable manager you are, so they too must be kept under some sort of control in order for your horse to feel his best and do his best. Picked up in pasture, worms are the greatest single parasite the horse can have. A sample of the horse's stools, taken to your vet, will show whether or not your horse has worms; and there are virtually none that do not. The vet can prescribe medication for the horse that you can give to him, or you can buy one of the many good commercial wormers on the market today. Either way, a strict regimin of worming should be included in your stable routine. Worming on a regular basis will keep the worm population in your horse's body at a minimum and keep him fit.

Trimming bot fly eggs off the legs of the horse, when you see them, and destroying them is also a way of keeping this parasite out of your horse's system. These eggs are small, yellow specks that cling to the hair of the horse's legs, usually around the ankles and cannon bones. Should the horse lick them off and ingest them, they will hatch inside his body. By destroying them when you see them, you cut down on the fly crop for next season and also keep them from getting into your horse's system.

Rodents are still another stable menace. Rats and mice eat, and run off with, hundreds of pounds of feed in their lifetime, if left unchecked. Rodents of this type also carry diseases that can be transmitted by their bite and their droppings. A menace to both animal and man, rodents should be delt with professionally. An exterminator is best for your rodent control. Poisons that you might put down for the rodents may be dropped by them instead of being taken away or eaten. If the poison is dropped where the horse can get at it, it could kill or injure him severely if he should eat it. Rodent removal is best left to the professionals. However, a cat or two around the barn is still a good idea for between extermination times. Just be sure to keep the cats out of the barn while the exterminator is there and for a number of days after he has done his work. Should the cat get a hold of a poisoned rat, he too may become ill or perish.

If you do not have cats and you want to remove as many of the rodents as you can yourself, then traps are the best idea, as long as they are out of reach of horses and children.

Mange, though not as prevalent as it used to be, due to better and cleaner stable conditions, is also a scourge of stables once it gets started. A parasite or louse that lives on or under the skin of the horse, mange not only causes discomfort, but disfigurement as well. The horse is plagued with a constant itch and will be scratching himself nearly all the time. His hair will fall out or be rubbed off in patches and he will look and feel miserable. The vet will tell you what kind of mange your horse has and will prescribe something for it. Be careful when handling a horse with mange, since some types are transmittable to man. Some types of lice can be passed from fowl to horses, so do not stable horses where fowl will have a chance to perch above them. Scratchings and dropped feathers can pass the lice to the horses, if they should happen to drop on the animals stabled below.

Thrush is yet another disease that horses can get, mostly from poorly kept stables. A fungus infection of the frog of the foot, thrush is easily treated in the early stages, but left unchecked it can severely damage the horse's feet. Wet and dirty stables are the most common causes of thrush today. Some horses are more prone to thrush than others, but most horses will come down with it if they are left in dirty stables for any length of time. Your vet will give you treatment for it or you can buy thrush treatment at any tack shop.

As you can see, there are many types of pests in the stable. All annoying, most are gotten rid of with good stable management.

6

FENCES AND PASTURE

I AM SURE YOU HAVE, at one time or another, driven by a field that was enclosed with a lovely white fence and seen a horse grazing contentedly behind it. Many of you have probably thought of how nice it looked and how much nicer than fields surrounded by barbed wire.

Well, there is a purpose to the different kinds of fences you see and many of those differences are quite important to stable management. One of the first things to remember when putting up fence for horses is never use barbed wire—NEVER, under no circumstances. Many a good horse has had to be put to sleep or has been crippled for life because of the ill use of barbed wire. A horse is an easily panicked animal. Once trapped somewhere, or by something, a horse will fight to free himself, sometimes to his detriment. He is unable to understand that if he remains calm he will not hurt himself. All he can know is that whatever has entangled him is painful and he must get away. Horses struggling in barbed wire only tangle themselves further, most often doing irreparable damage.

Cattle need the restraint of barbed wire, and because their temperament is different, as well as their hide, they do well enclosed with it. Horses do not. One single mistake around barbed wire may cost the horse his life, cripple him, or cause permanent scarring.

Riding around barbed wire is also a highly dangerous thing to do. It is very painful to get thrown into a strand or two of barbs.

I cannot stress enough the dangers and foolhardiness, of using barbed wire around horses. NEVER DO IT.

Regular fencing wire is better. It has no barbs on which the horse can get caught, it is strong, relatively inexpensive to use and put up, and it can be easily electrified if your horse is a fence breaker. The six- or twelve-

Nice example of white board fence.

volt charge used in fence electrifying will not injure the horse, but will teach him respect for the fence, and soon cure him of leaning on it. It may deter a fence jumper, too, if he has been allowed to touch the fence with his nose. I said allowed, not forced.

One thing to remember when electrifying fences. Use only a charger made specifically for animal fences. They can be purchased at any farm or stock store. Never wire your fence to the house current. You are condemning your horse to certain death, should he touch the fence, especially if the grass is wet or if he is shod, or both.

Regular wire fencing is easily put up and easily maintained. It is installed on metal posts, fastened to them by insulators, if you electrify, or wired to them if you don't. It needs no paint and can be quickly mended by splicing, if broken.

Another fence, which is prettier, is the board fence, spoken of in the beginning of the chapter. This fence does make a nicer appearance, but is harder to maintain. It needs painting, usually once a year, it is made of wood and will eventually rot, and it can be broken by kicks from playful horses. Too, wood is expensive and cannot be spliced when broken. The wooden posts, which are set in the ground, do not give like wire, but will rot off, usually at ground level. Often the rotting is not noticeable until a whole section suddenly goes down from a playful nudge from a horse.

Two strands of barbed wire. As brush grows up around it, it becomes difficult to see and even more dangerous to the horse.

If you have unlimited funds to spend on your horse project, then the best thing is post and rail fence, using locust posts. Locust is an extremely tough wood, resistant to rot, and the post and rail needs no paint and very little upkeep, except for occasional breakage.

A method of fencing seldom seen in the United States, but one which is fairly common in Europe and Great Britain, is the use of hedgerows. Usually blackthorn, whitehorn, or hawthorn, these thorny bushes are trained to grow into hedgerows. Since the branches naturally grow in a twisting manner, it is fairly simple to get them to interweave with each other as they grow. Once established, a hedgerow is nearly impenetrable. The interweaving of the branches gives the hedgerow strength to with-

A good example of a solid board gate and fence section. This is fine, as long as it is kept in good condition.

Beautiful fence of locust post and rail.

stand the pushing of the animals, and the sharp thorns on the branches deter most animals that try to push their way through.

Another alternative is to use pipe as fencing. Regular pipe, such as would be used for piping water in a house, can be secured to wooden posts with the aid of large brads or loops of heavy wire. The one drawback to this, however, is that there is no give to this kind of fencing and if a horse runs into it or tries to jump it and fails, then he is going to get hurt.

Livestock fencing, which is a type of wire made up into large squares, too, is objectionable. Too often, horses will put a hoof through one of the wire squares and get it caught, trying to pull it back. Again, the struggle will often cause serious injury.

To my mind, the best fencing for the everyday horseman is regular, single strand fence wire, electrified for best results.

One of the few examples of hedgerows used for fencing in the United States. Notice how they divide the pasture, without the use of either wire or board fencing.

Now, what about pasture? I feel that too little has been said about horse pasture. Too many questions have been unanswered, or answered unsatisfactorally, questions like: When is the best time to turn a horse out to pasture? Should he be clipped, should he be blanketed, or should he just be turned loose? Does a horse do better loose or tethered in a field? What kind of grass makes the best pasture? What kind of grasss are there for pasture?

To my mind, the best idea is to allow a horse to have as much pasture, each day, as possible, to give him some time in pasture every day or on as many days as he can safely be put out. This enables his system to adapt to the ever-changing conditions of the pasture, from spring green to winter cured.

However, this is not to be unsupervised pasturing. When putting a horse out to grass, if he has had none over winter, it is best to wait until the first rush of spring growing slows down. New grass is highly laxative and can scour a horse's bowels terribly. When putting him out on new grass, do so for only twenty minutes to half an hour for the first few days. Gradually increase his time over the next few weeks, until he is on pasture for the time he is going to be allotted daily. Watch his stools carefully. If they become too soft or runny or if they become fairly green in color, cut back the time he is on grass by half. You should notice his stools returning to normal within a day or two. If not, stop the pasture altogether for a week or two, then try again. Perhaps your horse cannot stand the early grass.

As the season continues, the grass looses some of the laxative properties and some very sensitive horses can now safely be put to pasture. If the horse is fed his usual ration of grain while on pasture, he will probably gain weight, especially if the grass is lush or if you have had a good growing year. If this is the case, and you don't want his weight to go up, cut down on his grain. Some horses don't need any grain while on good, rich summer pasture. However, it will be necessary to feed small amounts of grain if you want your horse to continue coming to the barn or when you call. Horses on pasture alone soon return to a seminatural state and become independent when not daily supervised. Also, if you feed supplements, you will have to feed some amount of grain in order to have something to mix the supplements into. Hay too may be cut down, the largest portion being reserved for feeding when the horse is going to be confined to the stable. If he is out at night, feed his largest portion of hay during the day, to keep him occupied. If he is out during the day, feed the largest portion while he is in at night.

A good way to feed hay when your horse is on short rations, for any reason, is to feed it from a hay net. The ration of hay is placed in the net, after being partially fluffed out, and the net is hung in the stall where the horse can easily reach it without either becoming entangled in it with head or feet and at a level where he need not reach up for it. This keeps bits and pieces from dropping into his eyes and nose. Hay nets slow the horse down when eating, since he has to pull the hay from the net before he can consume it. Hay nets are good for any horse that eats his hay too fast. When hanging the hay net, be sure that you do not hang it above the water bucket. You don't want his bucket full of bits of hay all the time, and neither does he.

When putting a horse to pasture, it is a good thing to begin cutting down on his oats a week or two before he is put out and gradually increasing his hay quantity. This allows his stomach and system to get used to the dietary change of less grain and more grassy food. It will help to lessen the laxative effects the grass may have on him.

If a horse is put out to pasture each day, he will go into fall with little weight change, as long as the pasture is good and his diet of other foods is closely watched. In late fall you should begin to notice his weight starting to drop a bit (earlier in other parts of the country), and now you should begin to increase his oat ration and drop the amount of hay he is getting. Gradually build up the oats and slowly cut down on the hay. You may want to leave the hay net for a few weeks, since he may be inclined to bolt his hay if he is hungry. After all, when the pasture begins to fall off he will find less and less to satisfy him. Remember to increase his grain gradually. You do not want to founder him or give him colic.

His ration should be watched carefully, seeing that he does not become bound up by the amount of extra grain he is receiving. A small amount of bran may be added to his ration of grain, if he appears to become constipated.

Winter pasture, especially in the midwest, if allowed to grow long and not grazed upon over summer, makes excellent pasture for a horse on days when the weather permits him to be outside. Pasture that has been allowed to stand ungrazed for a year is full of nutrients that are good for a horse.

Now, what about clipping and blankets for the field? If you have a horse that is being hunted in the winter, I would suggest both a hunter clip and a blanket for the field. Horses with heavy coats that hunt sweat off a great deal of weight and are prone to catching cold, since their hair is hard to dry, because of its length and thickness. This is not to say that all horses should be blanketed and clipped during the winter, only those who work hard during the winter months. The clipped coat allows the sweat to dry more rapidly after a hard sweat, as long as the horse is blanketed when not in use. When on pasture, a clipped horse should be covered. All other horses that are not being used in the winter should be allowed to grow a normal coat and be left unblanketed.

What about tying a horse in the field? This practice is called tethering. It is often used when there are no fences to confine the horse and the owner wants his horse to get pasture. This is alright IF your horse is used to it. Some horses are rope wise and used to grazing without stepping on the rope or having it wrap around their legs or otherwise entangling them. However, most horses like this are found out West and the number of truly rope wise, tetherable horses is fast disappearing. One reason is that horses are simply not used to work cattle and farm as they were before, at least not to as great an extent. More and more horses are being used as family horses and show horses, and while some western show horses need to know rope work, the majority of your everyday horses don't. Some horses, in fact, have never seen a rope, other than their lead shanks. It is best, unless you know your horse to be positivly rope wise, not to tether. Pasture under fence, if you can, or stay with him (with him on a lead) when he needs grass if you can't pasture under fence. Better that than a serious injury if he should become entangled in his tether and panic.

An example of what can happen when you graze horses on first year pasture. Notice the large bald patches where the grass has been pulled out by the roots.

Second year, permanent pasture. It is lush and green and ready for pasturing.

Planted pastures are mainly comprised of rye grass mixed with other grasses that have been selected for their hardiness and weatherability. This mixture of grasses is planted, and the best arrangement for growth is to allow the grass to grow for one complete season, after planting, without grazing anything on it. This permits the roots of the grass to gain a good hold and become established so that grazing will not disrupt them or tear them loose.

Grazing on first-year pasture is risky. If the young grass is torn loose you will have bald patches the next year, or will have to completely reseed and start over again. Once established, this type of pasture regrows, season after season. It is called permanent pasture. However, in order to do things right, you should have separate pastures, alternating as each successive pasture gets eaten down. If you have enough land, it is best to leave one pasture empty for a season and use another. Then the next year reverse the process. This allows the grass to rejuvenate. If your land does not permit this, then divide the pasture you do have and switch the horse from one pasture to the other as soon as the grass shows signs of definite wear.

Kentucky blue grass is used for pasture in Kentucky, Virginia, and several other states; however, much of the country has summers that are too hot for too long with too little rainful to sustain the blue grass.

The best idea is to find out the best pasture for your particular area

Here is an example of what will happen if the horses are not moved from one pasture to another. Notice how worn and eaten down this field is.

from your county agent or feed-supply dealer. There are no pat answers for pasture, since weather conditions vary greatly around the country. Knowing how to pasture and when, however, are a large part of being a good stable manager.

Another thing to consider when pasturing your horse is, does he have drinking water? It is necessary for him to be able to drink freely when he wants to while in pasture. The best possible thing is to have a free-running stream in his pasture, one that is not going to dry up during the summer heat when he needs it most. It is better than a still pond because most ponds have some places where the water backs up and these places are prime breeding and hatching places for mosquitoes. And mosquitoes carry diseases transmittable to the horse—diseases that can kill. If you

Marshes, like this one, are very poor pasture indeed.

Rocky pastures and pastures growing wild with weeds are bad terrain on which to pasture a horse.

have a pond, the best kind is one that has a constant flow of water, both in and out. This keeps the water circulating and does not permit the stillness needed for mosquito conditions.

Boggy or marshy pastures are also a poor risk, for the same reason. Anywhere that water stands or collects and is not moved soon becomes a breeding ground for insects. Too, marshy pasture often pulls shoes loose and is a prime place for the horse to slip, fall, or sprain himself. Bowed tendons are fairly common when horses are pastured in bogs or marshes. Foot problems also arise if the feet are not given ample time to dry out between pasturings.

A good pasture, under a suitable fence. The horse has room to both run and graze.

Rocky pasture, too, is a poor place to ask your horse to spend his days. Pastures that are full of rocks and stones are an invitation to bruised feet and twisted ligaments and tendons. All horses like to run when turned out to pasture, but doing so in a rocky pasture is courting disaster.

Ideally, pasture should be large enough to give the horse ample room to run as he sees fit, it should have a stream and the grass should be mature and well established in order to keep him well supplied with food. All of it should be contained under a suitable fence for your area and climatic conditions.

For your horse's safety in hunting season you might also post your property against hunting and trespassing. Place the signs in prominent

All property should be posted with large, easily read signs. If you do not wish trespassers on your property, then warn them away.

places, such as on your fence posts, and space them so that each sign is within sight of the other.

If possible, keep the horse in a pasture nearest the barn and during deer season keep him in the paddock or corral. It is for his protection.

Protect all of your livestock by posting signs of various types along the border of your property.

7

TO SHOE OR NOT TO SHOE

MANY PEOPLE, AND I was one up until just a few years ago, feel that a horse must be shod at all times to be in the best of condition and readiness for a ride or work. This is not so. There are times to shoe and there are times when it is not advisable to shoe, times when it is much better to leave the foot bare and in a natural state.

To begin with, let us look at the basic structure of the horse's hoof. It is nearly round and has supporting walls, a natural shock absorption quality called the frog, and a cupped sole. This is the condition of the hoof in its natural state. Man, however, has done all kinds of things to modify the hoof until, in some instances, it hardly resembles a hoof at all.

The foot of the wild horse or the horse in a natural state, will be trimmed by nature. Rocks act as files, which take off the long edges, and running on different terrain trims the sole and adds or removes moisture as the conditions vary. The domesticated horse, however, is usually kept in stone-free pastures, soft bedding, and is only given limited times and periods for running. Therefore, he is in constant need of something, other than nature, to keep his feet in trim and shape.

To start with, let us think of a horse who is in pasture most of the time and is ridden or worked little or not at all. Such is the lot of the broodmares of stud farms, whose sole purpose in life is to produce foals, either for racing, breeding, or showing. These mares have little or no need of shoes, since they stand on grass most of the day and are stabled at night. Also, a shoe would confine the feet and when the foal becomes large and ponderous before birth, a shoe would restrict too much the spreading of the feet, which compensates for the added weight of the

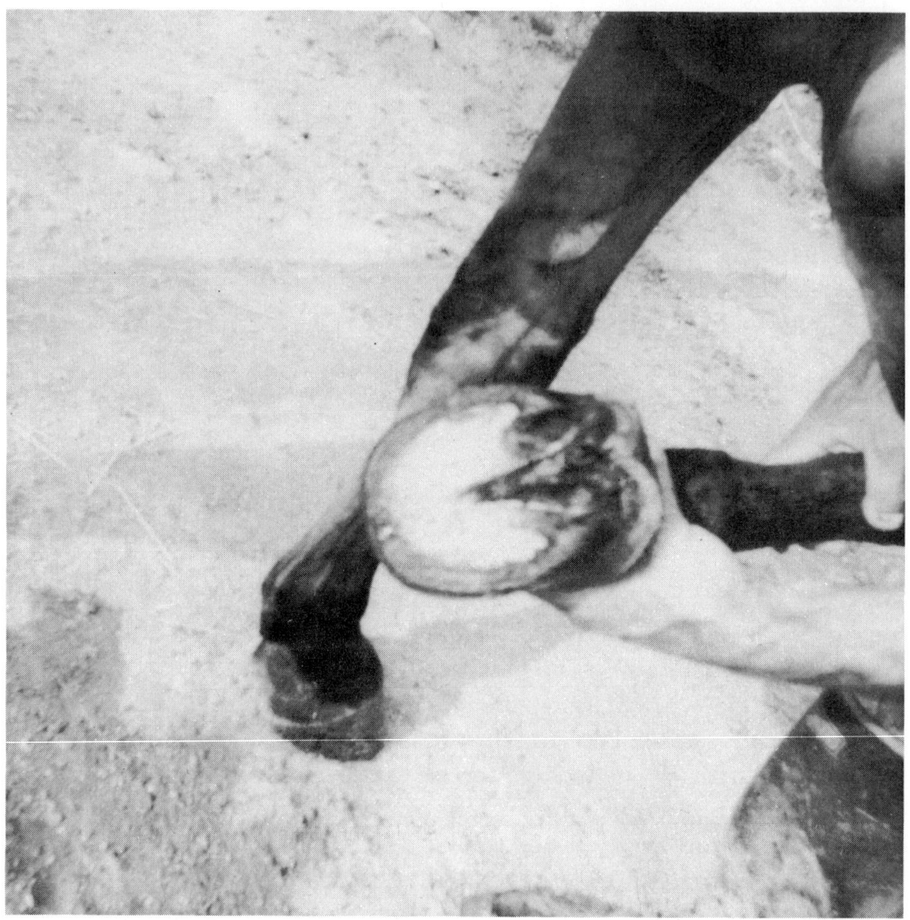

A beautiful example of the hoof in its natural state.

foal. Saddle horses, too, that have been bred should have their shoes pulled before the birth of the foal. This permits the foot, which has been confined, to relax, spread, and better support the horse's weight. The shoes should be pulled during the last three months, but no later than the last month, prior to birth. Having shoes pulled before the birth of the foal also lessens the chance of the mare hitting the foal with an iron-shod foot during and after the birth. It is a good idea to keep the shoes off while the foal is with the mare, so that there is no chance of her injuring him severely, should she accidentally kick or step on him during the time he is with her. If you cannot keep your mare unshod until the foal is weaned, it is a good idea to leave her unshod at least until the foal can fend fairly well for himself and is able to get out of the mare's way without effort.

Leaving the shoes off after birth also permits the feet to contract naturally and get used to being without the heavy weight of the foal. If the feet are allowed to regain their natural position there will be less strain on the legs and tendons of the mare when the time comes for reshoeing.

The feet, though not shod during pregnancy, should be kept trimmed by a farrier (blacksmith), and they should be watched for dryness and/or cracks and undue shelling off or peeling of the outer wall. This applies to any barefoot horse. The feet should be closely watched to see that they are not becoming too dry, getting cracked, or losing too large pieces of the wall. Some horses cannot go barefoot for a long period of time, especially if the ground they are on is excessively dry or hard. Often a hoof preparation will be needed to keep the hoof in condition by moisturizing it, to prevent cracking.

There are many foot and leg ailments that have been caused or brought on by faulty and often careless shoeing. Shoeing with shoes that are too small and do not fit, or by permitting the blacksmith to fit the foot entirely to the shoe, instead of fitting the shoe to the foot (corrective shoeing being the exception), all cause foot problems.

Shoeing too tightly in the heels is one of the prime causes of contracted heels, which may or may not be corrected by better shoeing. It also has a definite tendency to promote navicular disease, by compressing the blood vessels and nerve endings of the heels, thus cutting off the blood supply to the feet and killing the sensitivity of the foot. The loss of blood to the heels causes the degeneration of the bone. Contracted heels is the squeezing together of the heels of the foot, which often causes the heels to permanently grow close together in a confined position. This tendency can sometimes be reversed if caught early enough. Proper shoeing, with heels that are wide enough, or permitting the horse to go barefoot for a good amount of time, or having the horse shod with a shoe that has a spreader bar in the heels of the shoe, may alleviate the condition of contracted heels; however, sometimes the condition will not reverse itself. Sometimes the feet cannot be made to come back to natural growth pattern and the heels may be permanently squeezed together.

Corns can be another problem due to ill fitting shoes. Just as shoes that pinch or rub your feet can cause corns, so can shoes that rub or pinch your horse. And the results are the same. A painful corn will form where the irritation has occurred and nothing short of relieving the pressure, or in severe instances cutting, will alleviate the problem. Going barefoot will sometimes help in getting the corn to go away, but sometimes it will be necessary for the farrier to cut the corn away. Either way, you usually have a very lame horse on your hands—lame with the corn or lame after treatment. It is best to be sure the shoes fit the horse well and save yourself problems later.

A fairly common problem with shod horses is the tendency of the owner to permit the shoes to remain on the feet for too long a period of time, without having the shoes reset and the feet trimmed. Feet that have been shod and have gone too long without a trim and reset will often grow out over the front and sides of the shoes, which can cause severe damage to the wall of the hoof, by pulling the wall away from the live part of the hoof. Some hoofs also grow in such a way as to pull the nails down into the foot as the hoof grows. This can cause a serious problem when the time comes to remove the shoes, as there are no nail heads to cut off for easier removal of the shoes. It also may imbed the nail in the hoof, if the shoe is twisted while on the foot.

Feet that are not trimmed and reset often enough also have a tendency to develop cracks from the strain of trying to outgrow the shoes, and twisted tendons and strained muscles of the legs and sometimes shoulders, can occur if the feet are excessively long. If the feet are permitted to grow too long the horse has a tendency to become clumsy and trip and stumble and he may become unsafe as a saddle mount, until his feet are trimmed to normal lengths.

Excessively long feet, as seen on many show horses, such as Tennessee Walking horses and American Saddlebreds, are trained to grow to these extreme lengths in order to increase the high-stepping action required of them in shows. Blocks of wood or large leather pads are strapped to the feet to train them to grow in the long toed fashion for the shows, and the feet are trained like this nearly from birth. However, this is a very unnatural condition for horse's feet. Many of these horses develop trouble with their feet and legs as they age. The feet usually loose all of their natural shock absorption quality, as the frogs are unable to touch the ground, and many leg strains and tendon twists result from the highly elevated position of the feet due to the extremely long toes. It is a practice that I would like to see done away with or at least modified.

Shoeing, while a danger if improperly done, can also be one of the horseman's greatest assets. When working on rocky or hard ground a horse should definitely be shod. Many horses are worked on ground, with shoes, where it would be impossible to work them if they were unshod. The shoe protects the wall of the hoof from breaking when put under stress and it also keeps the sole from coming into too much contact with bad or rocky ground.

Sometimes a horse with badly worn down or broken hoofs can be brought around to usefulness again if the feet can be shod and the shoes kept on long enough to permit the foot time to regrow. Most trims and resets on a normal hoof should be done every six to eight weeks. Of course each horse varies, depending on how fast or slow the foot grows. That figure is average for an average hoof. However, I have known horses that have not had their shoes changed in seven months because the hoof was so badly worn down and damaged that there was hardly enough hoof left to permit the attachment of a shoe. It took that length of time for the feet to grow a sufficient amount to permit the shoes to be changed and the hoofs trimmed. This, however, was an exception.

There are many different types of shoes. To begin, let us look at the different types most commonly used. Most shoes are fullered shoes. Fullering is the groove around the underside of the shoe into which the nails are placed. This fullering allows the heads of the nails to be countersunk or driven flush with the striking surface of the shoe, thus keeping them out of the way and lessening the chance of the horse cutting himself with an exposed nail head.

Some shoes are seated out. This is a beveling of the inside edge of the front of the shoes on the striking surface. It lessens the weight of the toes of the shoes. This is needed by some horses that have a tendency to trip.

Some shoes have turned heels. This is a heel that has been made by the blacksmith by heating the tail of the shoe, pounding it, and rolling it over to form a heel.

Some shoes have toe clips or side clips. These are half-moon shaped clips, pounded out of the toes or sides of the shoes themselves, and they aid in keeping the shoes securely attached to the hoof, if the nails need to be placed low in the foot for any reason.

Shoes for corrective purposes may have trailer heels, the heels of the shoes being made excessively long and turned out. This aids some horses that forge or overreach and cut their front legs with their hind feet.

Some shoes have spreader bars in them, to aid in the correction of contracted heels.

There are also variations on calks and cleats that may be added to any shoe. There is borium, which is spot welded to the striking surface of the shoe for better grip on hard roads. There are mud cleats and ice cleats that may be attached to the toes and heels of shoes for the horse that pulls heavy loads or works on ice or in mud. There are also wide toe cleats for horses that work in deeply furrowed fields and need good traction.

There are virtually unlimited changes that can be made on shoes and in shoeing methods.

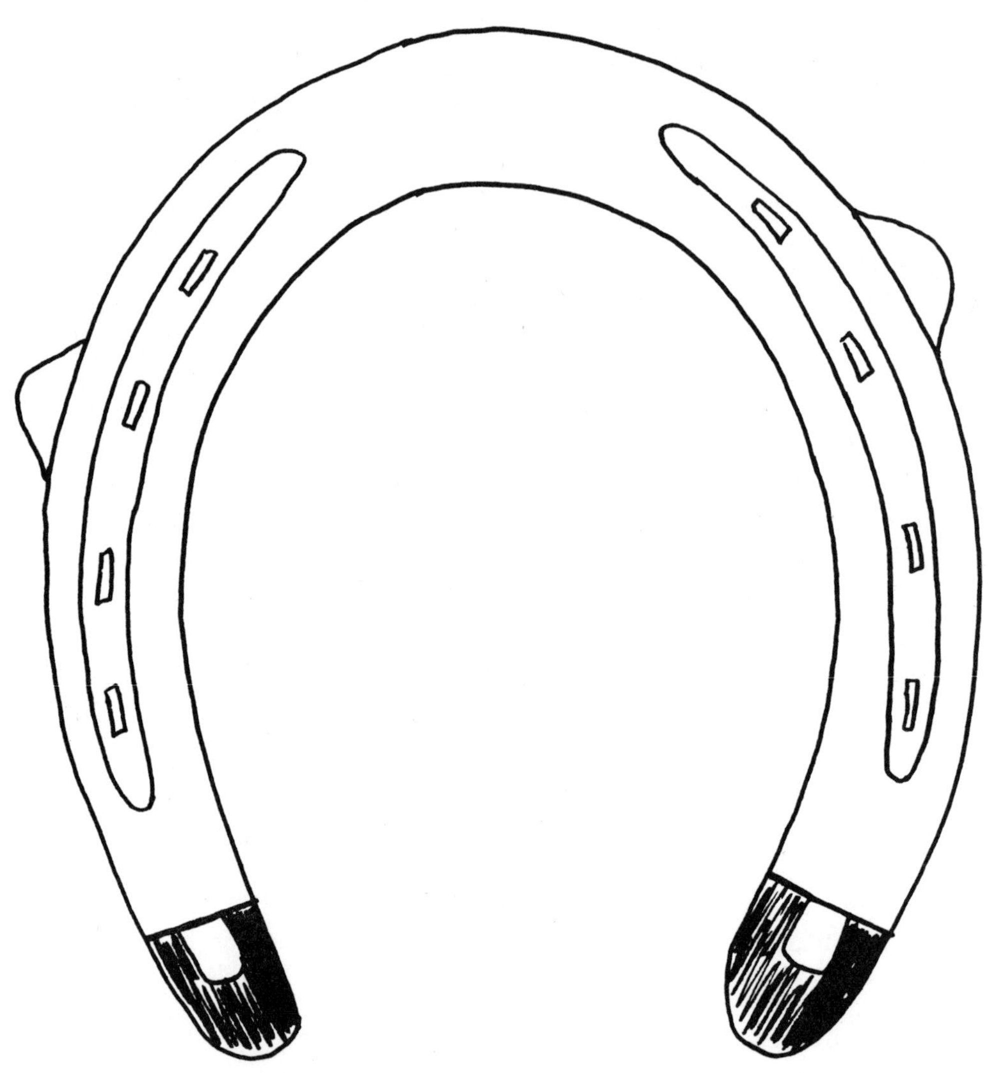

VII. Side clips and rolled heels.

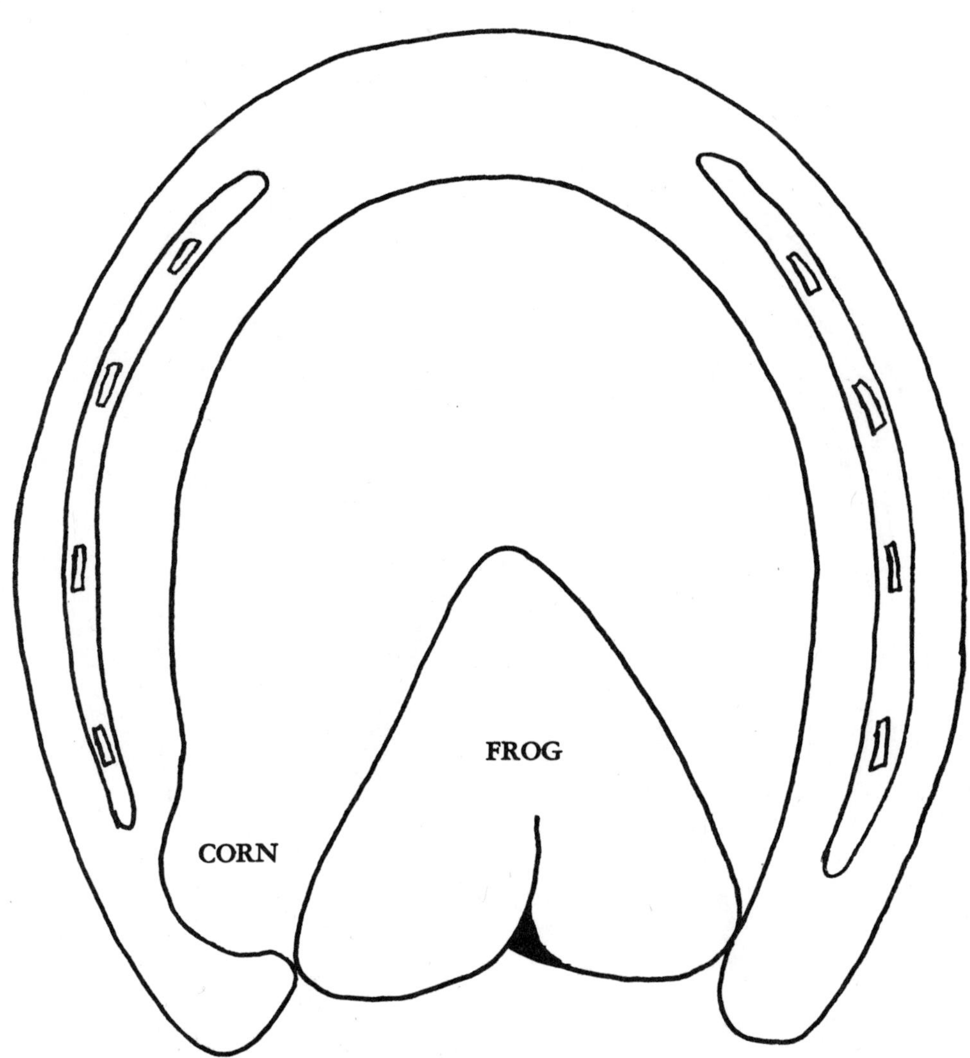

VIII. Shoe cut away to relieve corn.

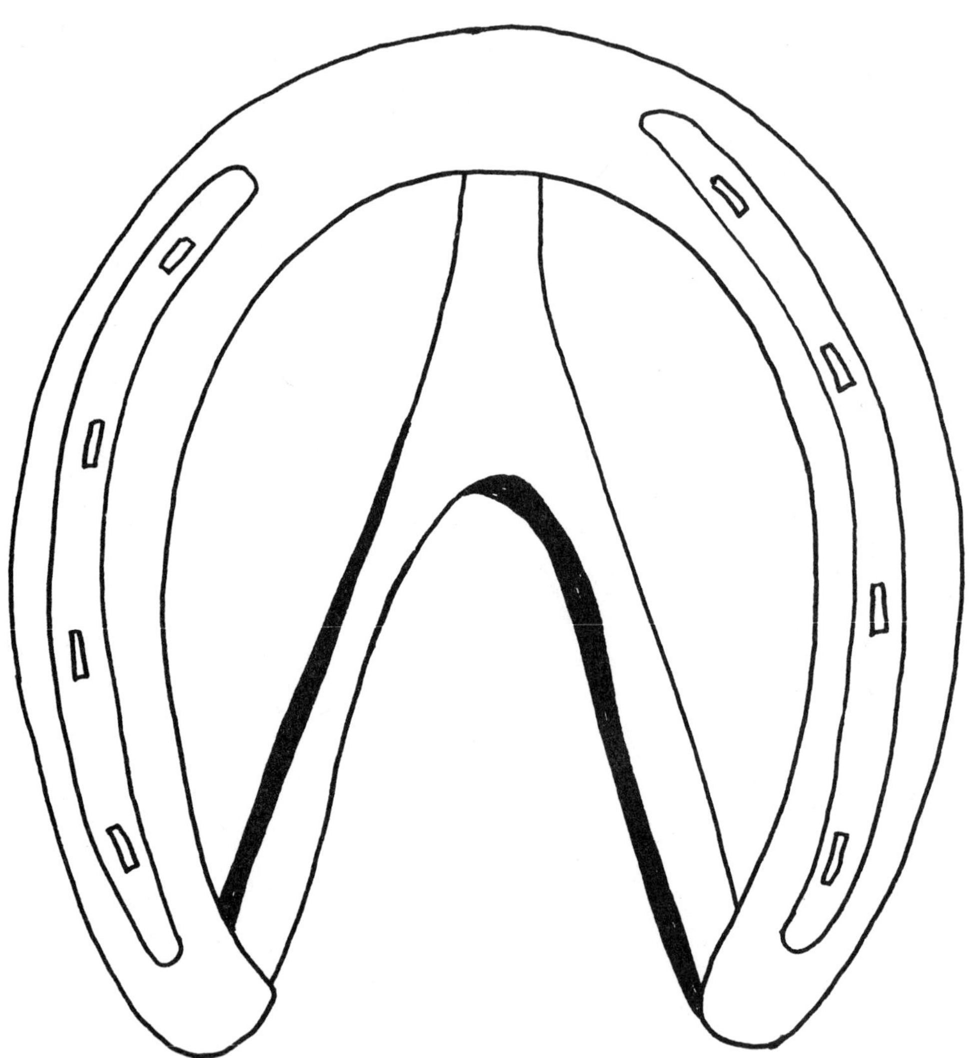

IX. Shoe with spreader spring—to cure contracted heels.

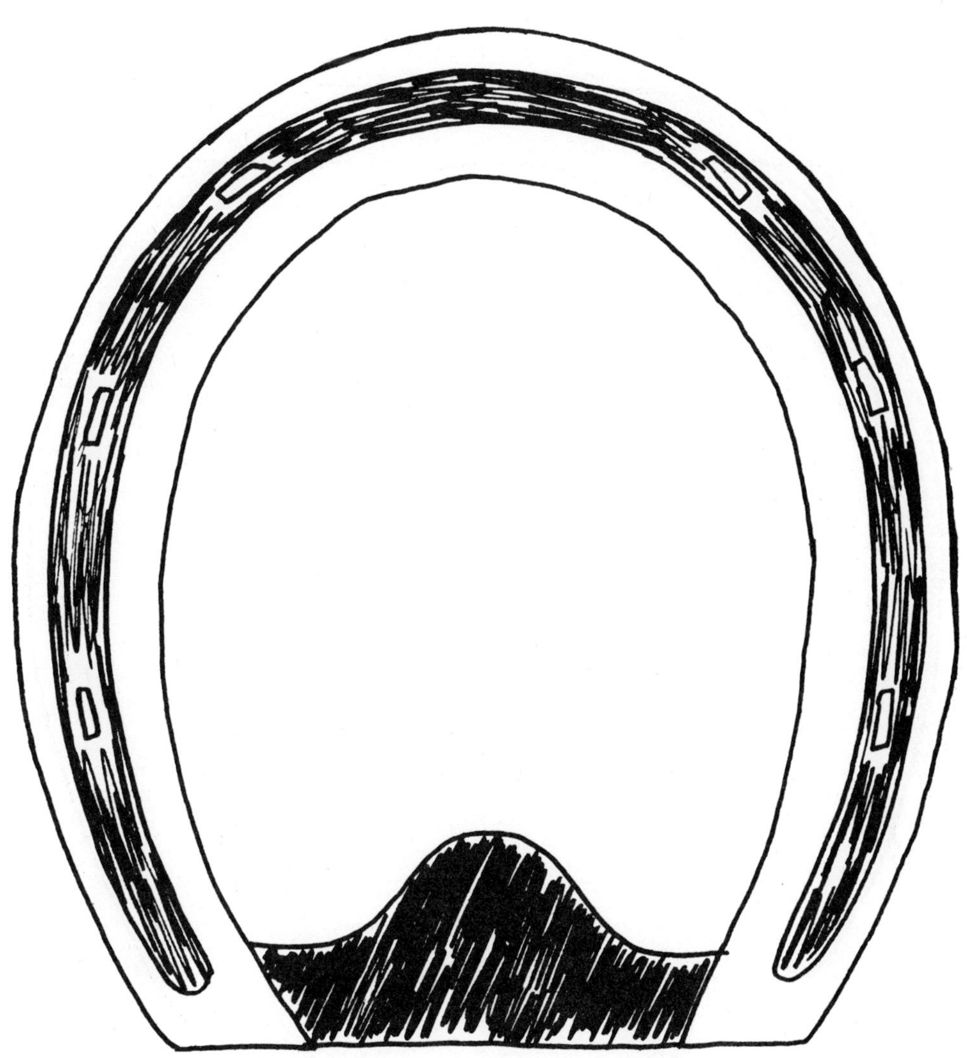

X. Bar shoe for support of weak heels.

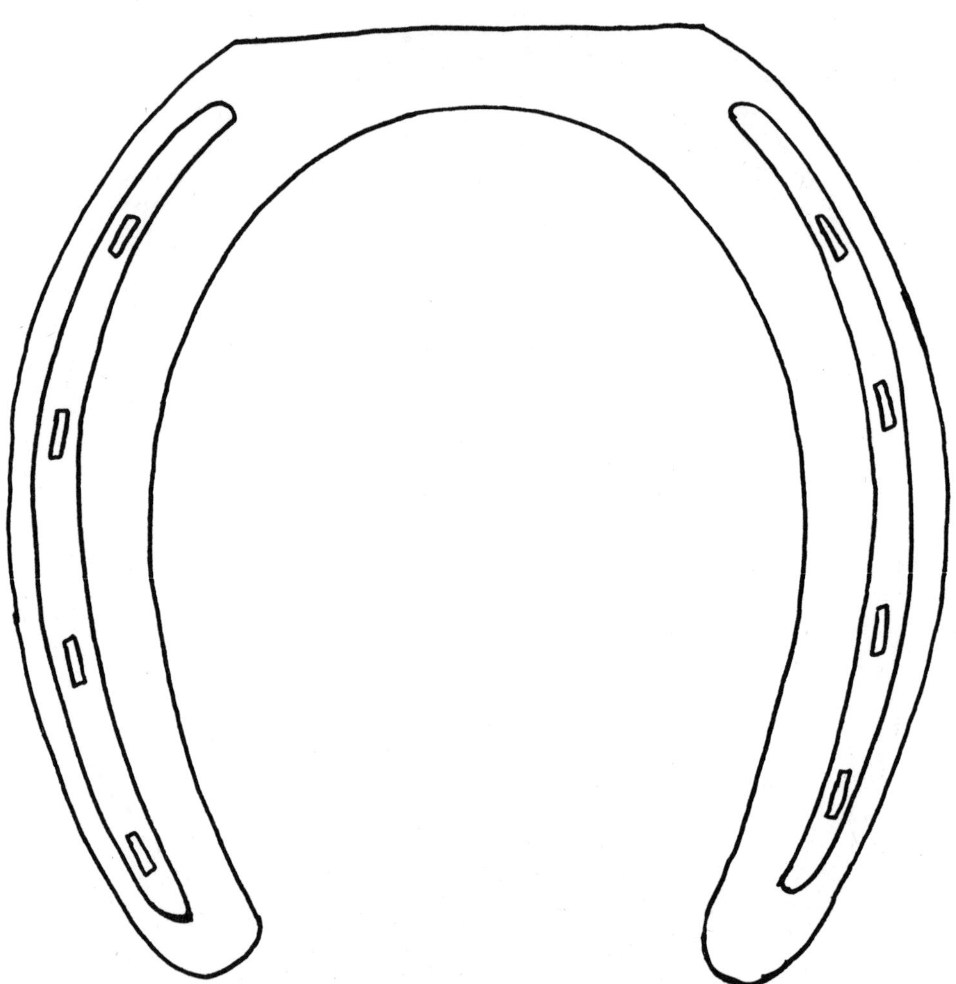

XI. Square or cut-off toe to cure or correct overreach.

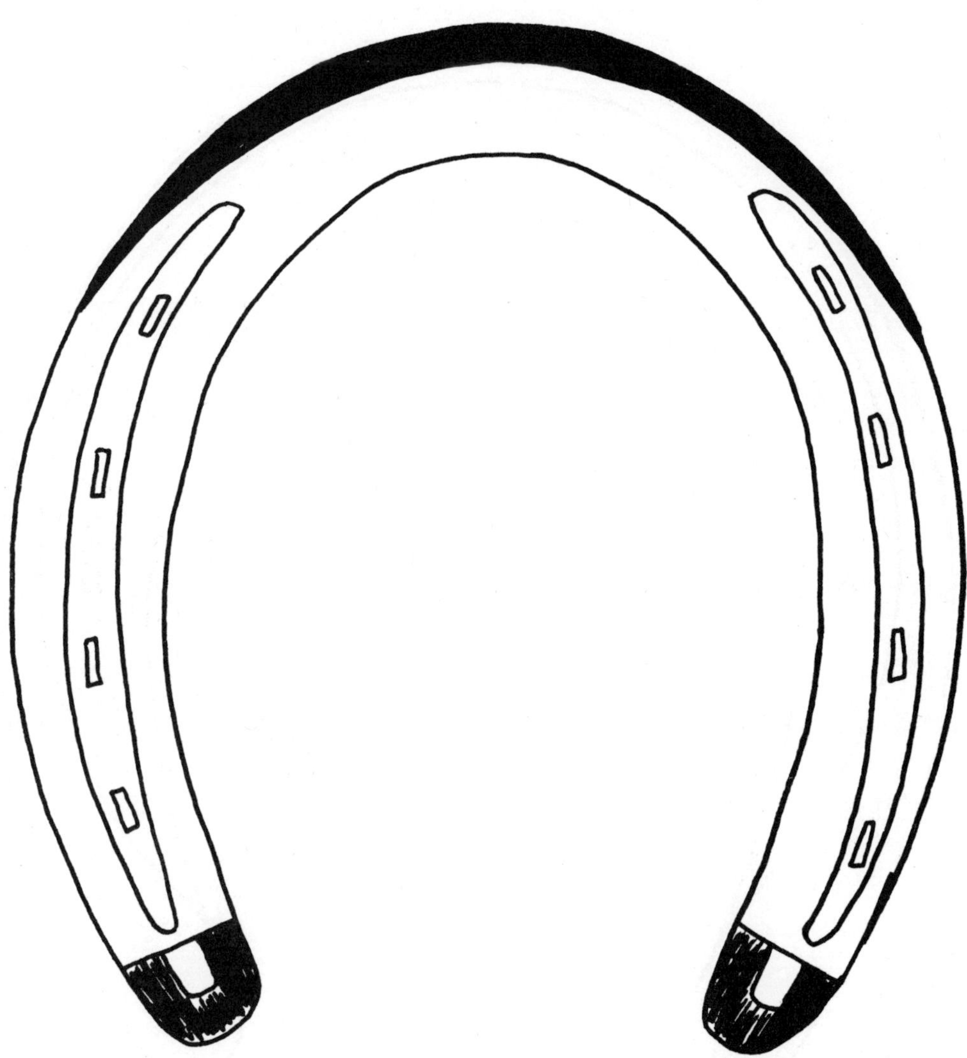

XII. Rolled toe for foundered foot or for stumbling turned heels.

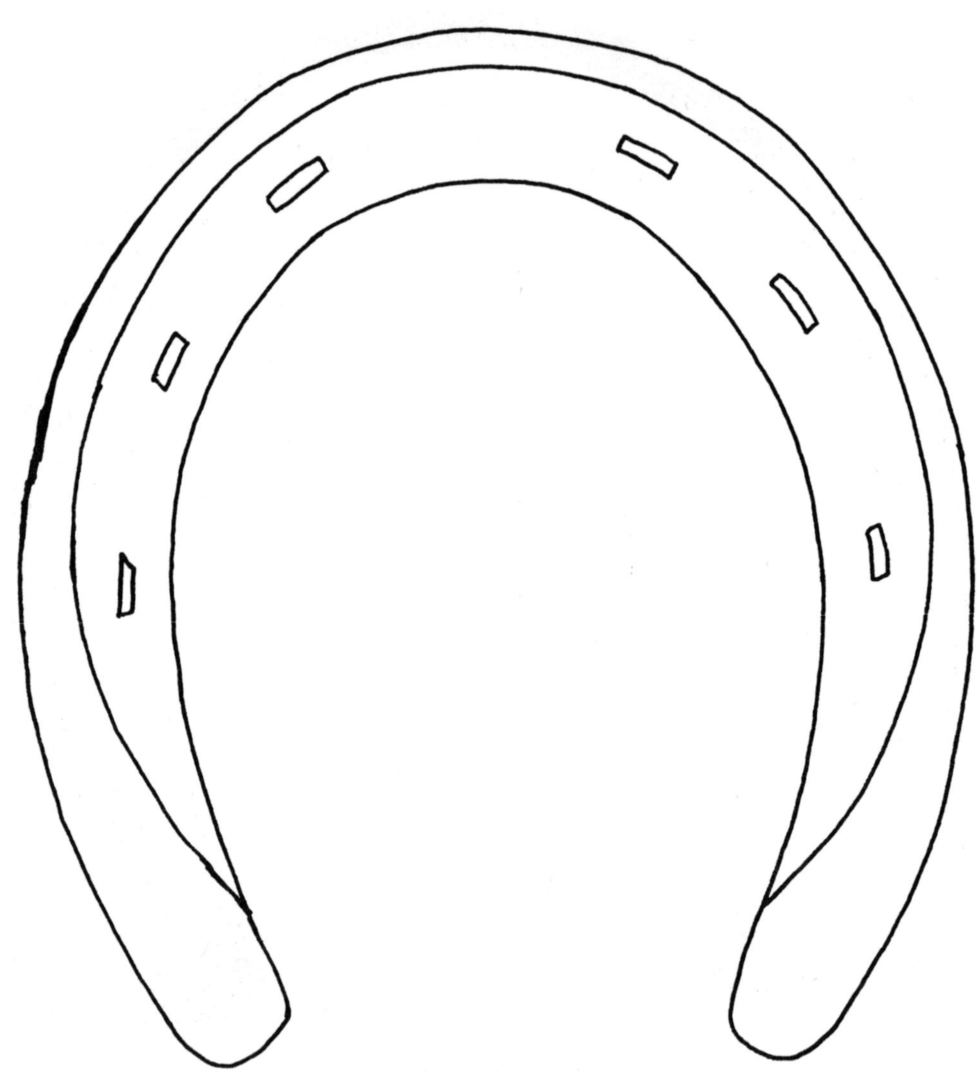

XIII. Concave shoe for extra-light weight.

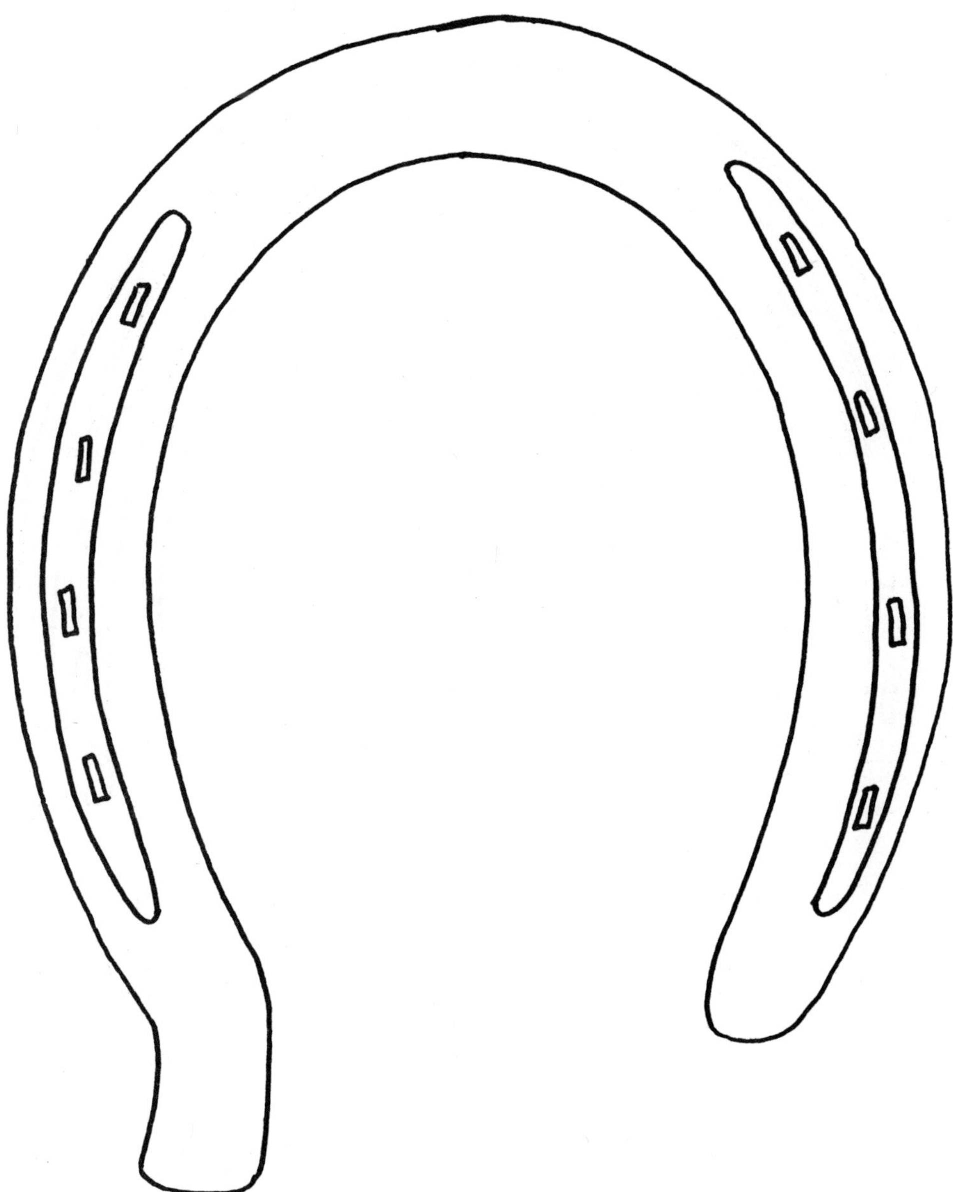

XIV. Shoe with trailer heel—may be on either heel of shoe—to cure forging and/or splayed foot.

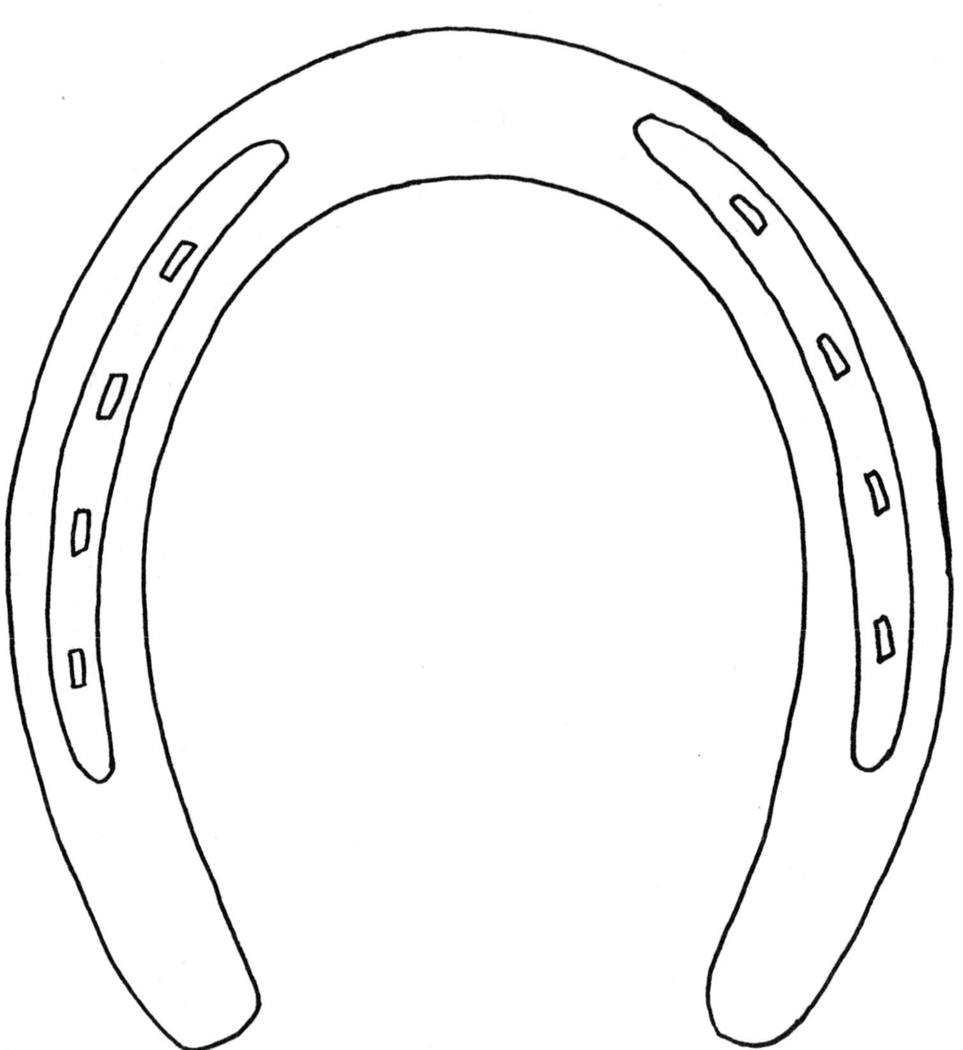

XV. Plain fullered shoe.

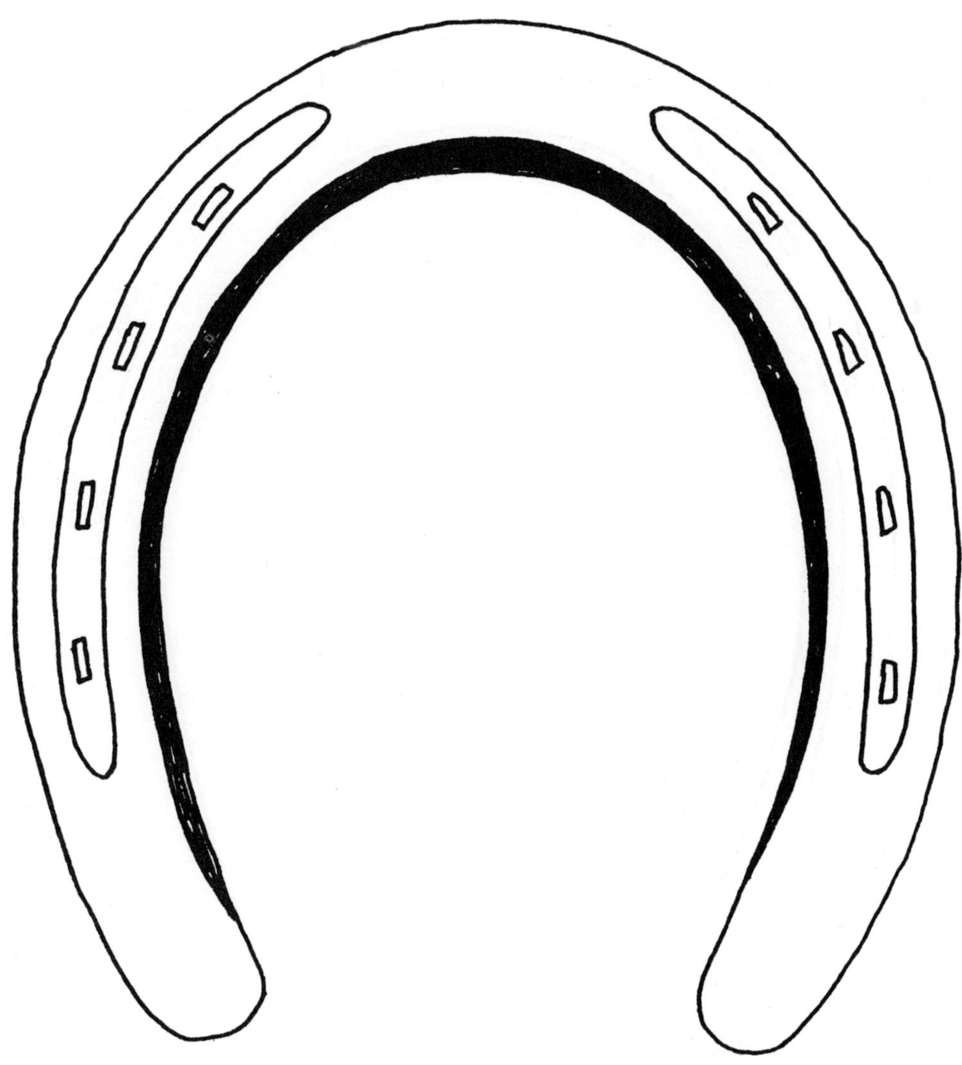

XVI. Fullered and seated out.

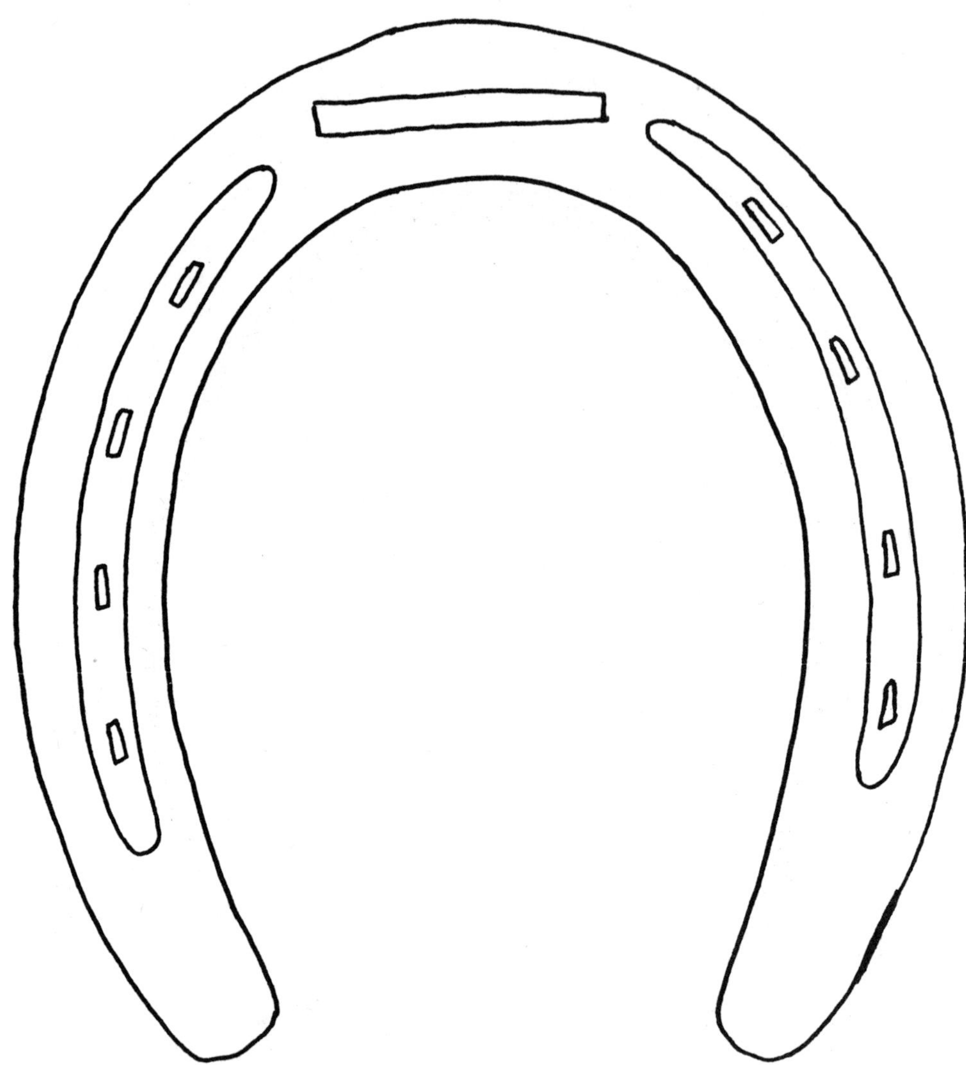

XVII. Toe bar cleat for use on carriage horse.

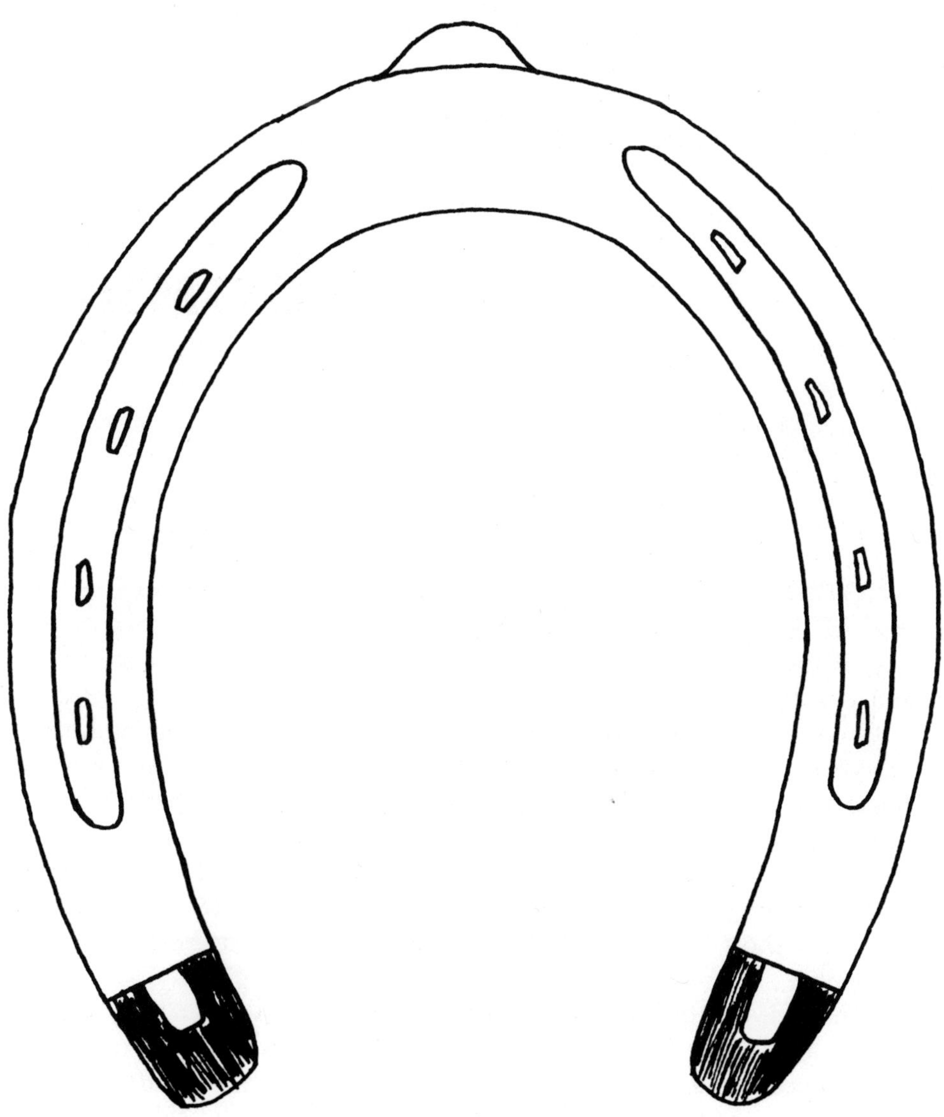

XVIII. Turned heel and toe clip.

8

GROOMING

ANOTHER IMPORTANT ASPECT of stable management is grooming. All horses should have a good, thorough grooming at least once a day. If the horse is ridden during the day, he should be groomed before and after his ride.

Grooming is as essential to the horse as combing, brushing, and washing your hair is to you. It is a part of basic hygiene and the only way the horse stays clean. To him, a good grooming is like a daily bath. It cleans him, refreshes him, and relaxes him, as well as toning his skin.

A full grooming is just what the name implies—a complete grooming, from start to finish. A short grooming, such as a brush down before a ride, or a short brushing off after a ride, is called quartering or quarter grooming.

A horse is usually fed in the morning, groomed fully after he is finished eating, then either allowed to stand in the stall, is turned out to pasture, or is ridden. If he is ridden then he should be brushed off after he has cooled down, and his feet should be picked out. This after ride brush and pick out is quartering.

Never brush or groom a horse while he is feeding. It is disturbing to him. Like trying to comb, brush, or wash your hair while eating at the table. You can't do either well or in comfort.

As with any other part of stable management, grooming has a right way of being done. There are also special tools for grooming. A complete grooming kit contains the following items:

120

GROOMING KIT

Rubber or Mud Curry	Pastern Brush
Shedding Blade	Mane Drag or Comb
Dandy Brush	Hoof Pick
Face Brush	Stable Rubber
Finishing Brush	Dock Sponge

Let's go on to the grooming itself. The first thing to do is to remove the horse from his stall, if possible, and take him to the cross-tie area. Remove his blanket, if he is wearing one, by folding it back from the front. Fold it back over his haunches, in half, lift it off him by the center of the fold and the edges that are together. This will cause the blanket to be folded in quarters, with the side that goes against the horse on the outside.

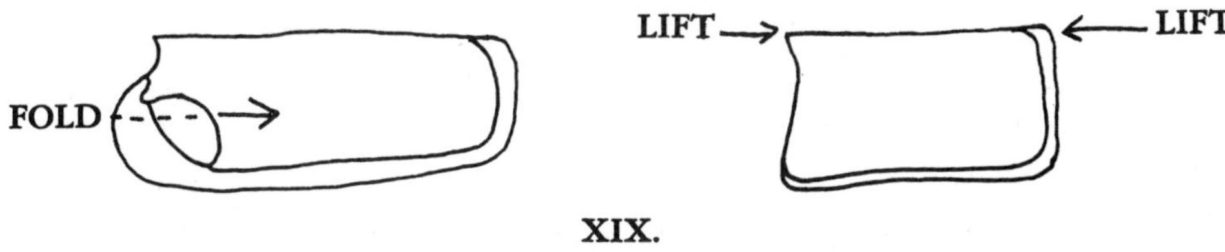

It is much easier to replace a blanket on the horse when folded this way. However, do not think that blanketing a horse is a replacement for grooming.

After the horse is cross-tied, take the mud or rubber curry and remove any matted mud or manure with circling motions. Never use the mud curry below the knees, hocks, or on any parts of the horse where bones are close to the surface of the skin. Use swirling motions, with medium pressure, less if your horse is sensitive. After removing the mud you should take the shedding blade and gently comb the way the hair lies to remove any dead and loose hair your rubber curry has dislodged. Again, do not use this tool below the knees, hocks or on any part where you might come in contact with near surface bones. After removing the dead hair, you should next use the Dandy brush. The Dandy looks much like a large scrub brush and has long, coarse bristles. Beginning at the poll, brush toward the rear of the horse, the way the hair grows, using short, firm strokes. Clean the horse thoroughly all over with the Dandy brush, except for the face. The pressure used with the Dandy should vary, depending upon which part of the horse you are grooming. The neck and flanks may be brushed very firmly, as may the back; however, the loins, girth, belly, and legs may need less pressure, depending upon the sensitivity of your

A complete grooming kit. Clockwise: Shedding blade, mud curry, Dandy brush, face brush, finishing brush (above), pastern brush, mane combs (2 types), hoof pick, dock sponge, and stable rubbers.

particular horse. Watch his ears, since they will indicate if the pressure is alright or too firm. If the ears go back and down you are brushing too hard and should ease off.

After having brushed the horse with the Dandy brush, the next brush to use is the face brush. This is an extra brush in most kits, but should be included in each one. It resembles a small Dandy brush, but is much softer. Brush the face and ears with it, always brushing in the direction the hair grows. Shield the eyes and nostrils with the hand when brushing the face, in order to prevent dust and dirt from entering them. Brush the ears with the face brush, brushing from the base of the ear to the tip. Be sure to brush the inside of the ear as well, but do it gently. Many a horse has become shy of having his ears handled because of rough treatment them. This is one reason a face brush is good for ears. It is soft and gentle to them.

After the face brush, the finishing brush is next. Again, brush the entire horse with it, this time with longer strokes, in order to lay all the hair flat. The fine, closely set bristles of the finishing brush slick the hair down in one direction, removing fine dirt and dust particles and adding a shine to the coat.

Next, use a mane drag or comb for mane and tail, since brushes tend to split hair ends. Comb mane and tail thoroughly, small amounts at a time, until the hair is free of tangles and hangs smoothly. If the tail or mane have burrs in them, remove the burrs by hand before brushing by pulling the hair away from the burr, not by pulling the burr free of the hair. Remove very small amounts of hair at a time in order not to split or break the hair.

Next is another tool that is often not included in grooming kits. It is called a pastern brush. Resembling a tiny scrubbing brush, it is used, gently, to clean the coronet area and heels of the hoof. Mud may easily be brushed free of the heels and coronet area with this small brush and should you, for any reason, leave the fetlock hair on your horse, it is great for removing caked mud from the pastern area.

The hoof pick is the next item to be used. This is a small, steel hook, used in cleaning out the horse's feet. Holding up the hoof, place the point of the pick at the sides of the frog and draw it forward toward the toe of the hoof. Be careful not to press too hard or too deeply. Scrape the foot clean of all matter and manure. Always be sure to clean from heel to toe. NEVER draw the pick from the toe to the heel of the hoof. By doing so you drive any matter picked up by the hoof pick into the sensitive skin of the heels, and you may bruise the frog too. Clean all four hoofs in this manner, from heel to toe.

After you have finished with the hoof pick you should slick the horse up, all over, with the stable rubber. This is any napless cloth, although linen is best, run over the horse's body in the direction that the hair lies. It produces a very high sheen on his coat.

The last article of use is the dock sponge. A small sponge, the dock sponge is wrung out in warm water, and the dock area of the horse's tail should be washed, then dried.

Your horse is now ready for his ride, for pasture, or just about anything.

You like to be clean and so does your horse. Be sure he is kept clean and you will be a better master, as well as stable manager.

9

TACK

AS WITH ANY PART of stable management, there are good and bad methods of storing tack. Naturally, the better care you take of your tack and the more efficiently you keep it, the better stable manager you will be. Poorly kept tack results in broken tack, moldy tack, and usually ends up costing quite a bit of money in replacement and repairs. Well kept tack, on the other hand, is much less prone to breakage, tearing, or rotting, and will give you years of good service. Let us see just what constitutes good and poor tack care.

First of all, when you buy tack the number one thing to do is to clean and oil it. Cleaning and oiling new tack begins the softening process that is needed in breaking in the new leather. New leather is usually stiff and needs to be lubricated or softened, in order to be able to begin to take on the shape of your seat, legs, and hands, as well as to begin to conform to the shape of your horse's body. Thorough cleaning and oiling after each use quickens the process of breaking in and also keeps the leather flexable, for easier contact with the horse during the ride.

When storing tack it is best to have some sort of individual racks for saddles and bridles. Saw horses that have been thoroughly padded make good saddle racks. The saddles sit firmly upon them and are easily removed for use.

Pieces of inch-thick pipe may also be used for saddle racks. The pipes are attached to the wall with large metal brads and are fastened together with elbows to form a right angle. The saddle is then placed upon the pipe that protrudes from the wall. When not in use, the pipes can be swung back, flat against the wall, out of the way. These types of swing

A fine example of storing tack. Each saddle and bridle has its own rack and the saddle pads are neatly arranged on top of each saddle.

flat racks eliminate the danger of someone walking into the rack ends when they are empty. Also, they take up less space than do stationary racks. However, some racks can be obtained with wheels or casters on them for ease of movement.

Too, if you don't want to go to a lot of expense for saddle racks, old (or new) towel racks (the kind that stand by themselves) make wonderful, easily moved, saddle racks. They need no padding or additional work done on them, most have a small shelf underneath, and they can be taken apart and put back together quickly. They pack flat in small spaces and are great to take along to horse shows, since they are dismantled and reassembled so readily.

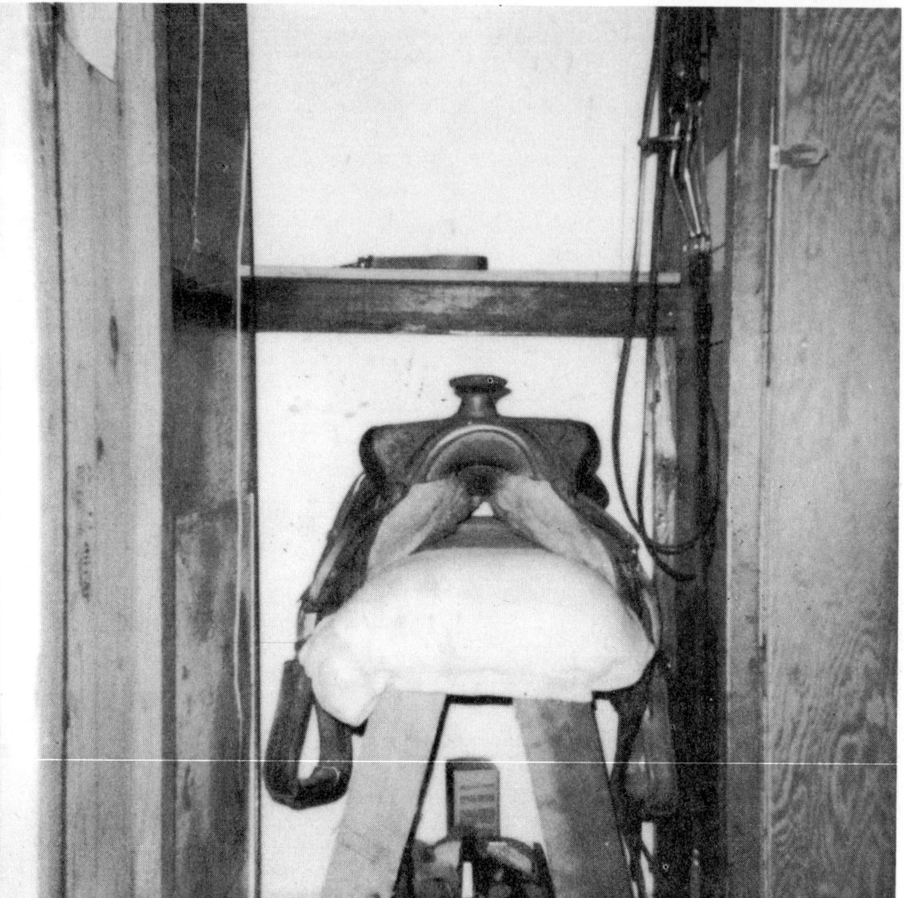

A well padded saw horse provides a solid rack for a Western saddle.

A separate room in the barn is the best way to keep your tack. It is
kept handy, yet out of the way things. Saddles collect dust readily and
storing them in a separate room cuts down on some of the dust, but not
all. You may want to make saddle covers for your tack or simply throw
a piece of an old sheet over your saddles. Another way to keep your tack
free of dust is to place each saddle pad upside down on the top of the
saddle after each use. That way each saddle and pad are together the way
they belong. There is no chance of delaying a ride, lesson, or training ses-
sion by searching for a missing pad. Simply brush the pad off, before and
after the ride, place it on top of the saddle for storage, and you are all
set. Placing the pad upside down allows the air to dry the part of the pad
that was against the horse's body, thus keeping the wet off the saddle,

drying the pad, and yet keeping both saddle and pad together and ready for instant reuse.

Bridles should be hung from racks, one bridle to the rack, and if you have the room, you can and should do the same thing with the halters. Each saddle, bridle, and halter should have the horses name on either the piece of equipment itself or on the rack that holds it. When returned to the tack room, each piece of equipment should be replaced on the rack designated for it and on no other.

Blankets may be stored in tack chests when not in use, though be sure they are thoroughly dry before you store them. Blankets, like anything else, will mold and mildew if put away damp. Blankets should always be brushed clean before storing.

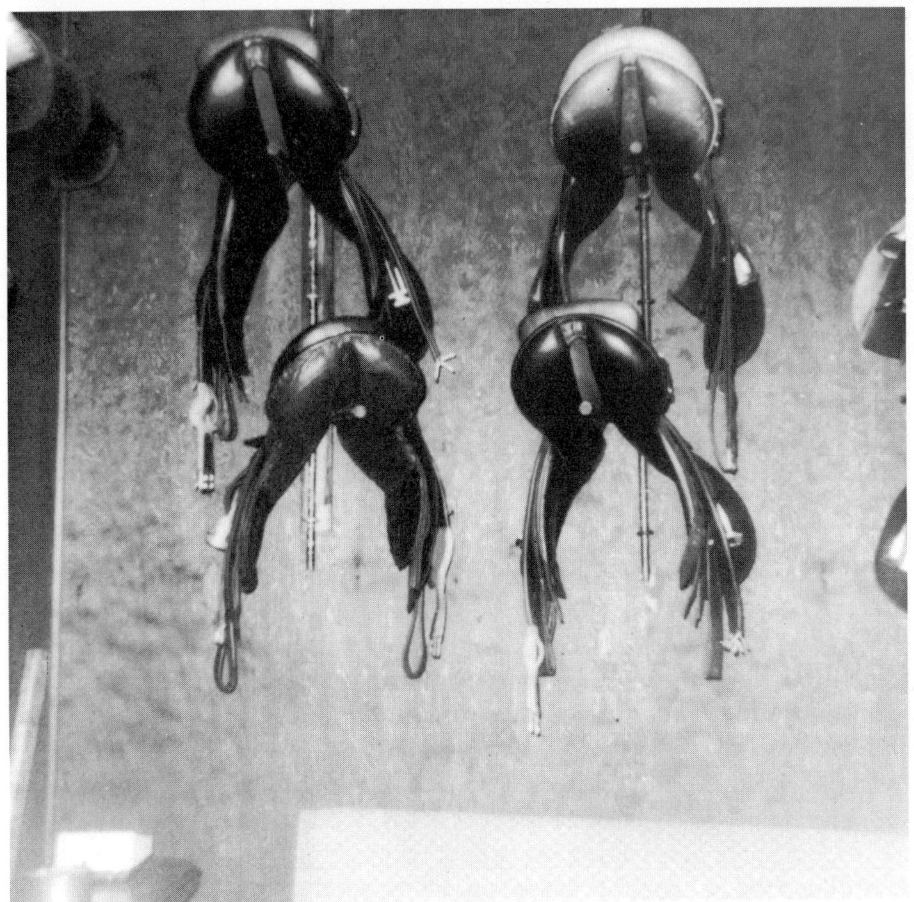

Right angles of pipe store saddle with a minimum use of space.

Here, bridles are each hung from separate racks, each bearing the name of the horse for which each bridle is to be used.

Saddles should be thoroughly aired and cleaned, then allowed to dry after each use and before they are put away. Standing the saddle up on its pommel and allowing it to dry is one way. Another is to place the saddle over the top rail of a fence and allow it to dry after use and cleaning. However, never dry tack in strong, direct sun for too long a time. Extreme heat is very hard and drying on leather.

Tack should never be allowed to remain uncleaned for more than a day after use. The sweat that is on it, as well as the dirt, will dry, cake, and harden onto the leather and make it extremely difficult to clean off later. Too, the salt in the horse's sweat is very drying and hard on the leather. Saddles should be racked, even if the cleaning has to be put off until later. They should never be tossed down in a pile or deposited willy-nilly. Bridles too should be hung up until cleaning time. It is easy to catch a foot or leg in a bridle rein and too easy to stumble over a saddle or stirrup iron lying on the floor or in the aisle.

When cleaning tack, you should rack it on a rack that is used just for tack cleaning. It should not be cleaned on its own rack, since racks that are used for cleaning soon collect soap and sweat scum. You don't want your saddle resting on that.

To begin cleaning your tack, first remove the irons and leathers. Open the safety bars of the saddle. Remove the irons from the leathers and put both aside for later cleaning. Now, take a tack cleaning sponge and wring it out in warm water. Next, rub it across your cake or tin of saddle soap

Three saddles set out to dry after a ride.

until you form a lather. Your sponge should be wet enough to do this, but should not be dripping. Next, apply the lather to the saddle and work it into the leather with quick, rounded strokes. Clean the entire saddle in this manner, washing out the sponge after each application and resoaping it each time. Be sure to clean around all stitching thoroughly and around all dees, billets, and the safety bars of the leathers. Be sure you clean the gullet of the saddle thoroughly and do not allow any sweat or salt to remain on the underside padding where the saddle rests against the horse's back.

Buff the saddle off, all over, with a clean cloth and set it aside to finish drying. Next, take the stirrup leathers and proceed to clean them in the

Note the proper way of placing the saddle in drying position, on the front of the pommel.

Here again we see a saddle drying. Though it will not dry quite as quickly as setting it up on the pommel, it will be out of the way on the fence.

same manner as you cleaned the saddle, being sure to clean well around the stitching of the buckle. Buff them and put them with the saddle, but do not return them to it.

Next, take your stirrup irons and clean them well with any of the commercial chrome polishes that are on the market. If they are extremely dirty or mud caked, wash them first with soap and water, dry them, and then polish them with the chrome cleaner. Be sure to buff them off well, removing any excess cleaner from them. Now, place them back onto the leathers, place the leathers back on the saddle, close your safetys, and rack your saddle. Be sure to run your irons up properly before leaving

Saddles, bridles, and brushes should never be left in this condition.

the tack room. Leathers not run up against the bars are a nuisance and a hazard: a nuisance in the fact that you must do it before you can saddle your horse for his next ride; a hazard because if they are not run up and you attempt to saddle up, you or someone else, may get hit in the head with one of the irons during the tacking up process.

Bridles should be cleaned thoroughly, too. If you haven't the time to completely strip down your bridle after each ride to clean it, then at least clean the reins and cheek and crown pieces after each use, especially around the buckles and stitching. Then, once a week each bridle that has been used should be taken apart completely and given a thorough cleaning and buffing. Leather that remains bent in one position for a long time, such as where reins connect to bits, bits to cheek pieces, etc., soon develops a tendency to dry out completely and quickly. Cracks, too, soon develop where leather is bent for long periods of time. Many of the natural oils in the leather escape from this frequent bending. Be very sure to keep all bent parts, of both bridle and saddle, well cleaned and oiled, in order to compensate for this natural oil loss and dryness.

Bits, too, should be washed and the bit joints that fit at the corners of the horse's mouth should be oiled, then wiped dry. Washing the bits takes off all old, dried saliva and bits of dried grasses or grains that may have collected on them. Careful oiling of the outer bit joints keeps them moving freely so that they don't bind and cut or pinch the sides of the horse's mouth. Never oil a bit joint, such as the joints of a snaffle, that goes into

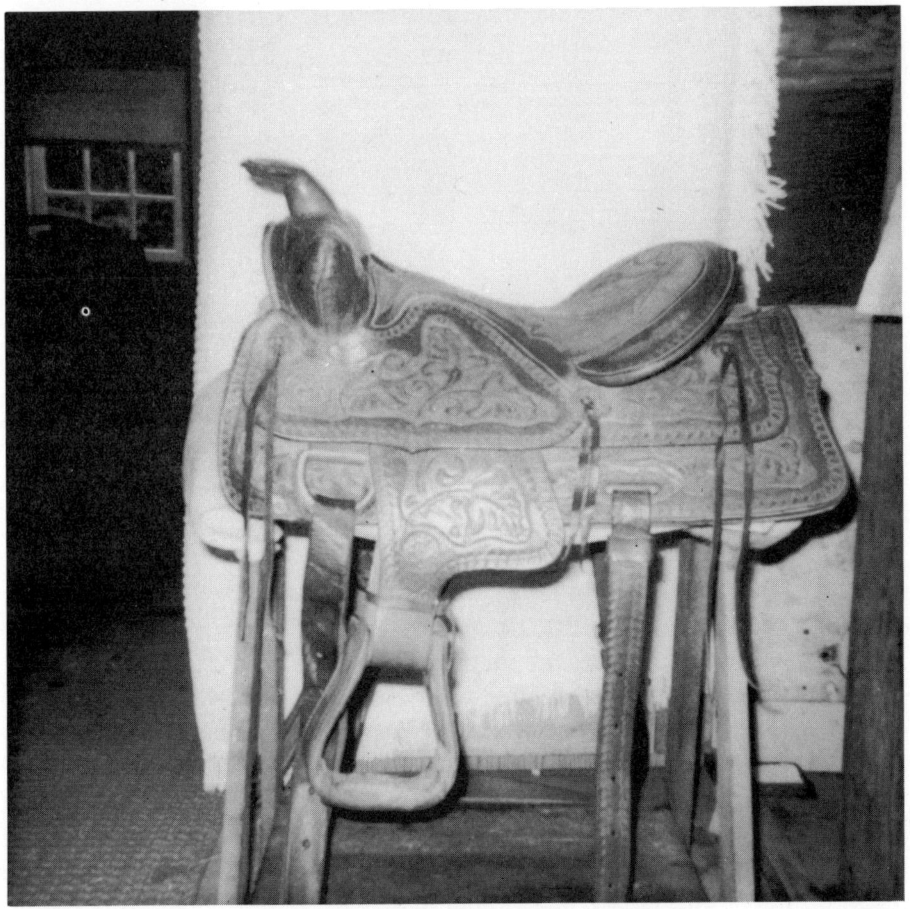

You should have one special rack that is used just for cleaning of tack. It should, besides being at a comfortable height, be firm and solid.

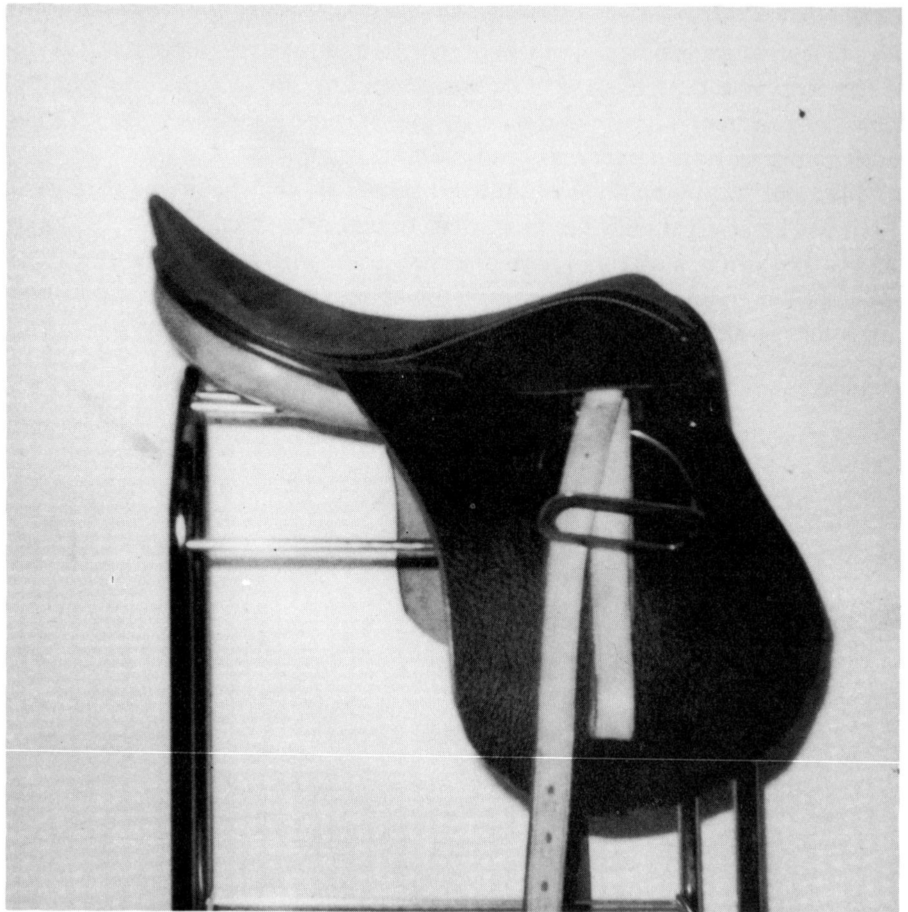

English saddle with the irons properly run up for storage.

the horse's mouth. He won't like the taste and may give you a problem over it. It may make him refuse being bitted in the future.

Curb chains should not be neglected in the cleaning, and may be polished off with chrome cleaner when you are finished. Bit rings and shanks may also be finished off with chrome cleaner, but again, be sure not to get the cleaner on the bit mouthpiece itself.

Saddle pads should be removed from under the saddle and should be brushed out every day, after they have dried. This removes any sweat or salt that has been deposited on them. They may then be turned upside down on top of the saddle for storage. Any pads that are washable should be washed at least once a month.

If you have old tack and you are trying to restore it to usefulness, you might try working some heated neatsfoot oil into it. Heating the oil helps the leather to absorb it. Make the oil warm enough to promote maximum absorption, but not too hot. It is the proper temperature when you can tolerate having it in your hands in comfort. And, for best results, apply it with your hands. Applying the oil with a sponge is less messy, but hand application is better and more thorough. The heat of your hands aids in keeping the oil warm while you apply it and the longer the oil is kept warm, the better it will work. As you apply the oil, work the leather back and forth, bending it gently to and from you, but being careful not to crack it. While the leather is bent, rub the oil into the bent surface, then allow the leather to straighten out. Apply the heated oil to all surfaces and allow the piece being worked on to dry by itself. Do not buff it off. Old leather will often need more then one application, twenty-four hours apart, before any noticeable results occur. Once the leather becomes more pliable, the neatsfoot treatment may be stopped and saddle soaping may be started. The only drawback in using neatsfoot alone is that it sometimes encourages mold growth or mildew; however, it makes an excellent starter for restoring old leather to usefulness.

Regardless of whether the equipment is old or new, its care is highly important among stable management duties.

10

TRAILERING OR HAULING

THE MOST IMPORTANT THING in hauling or trailering your horse is safety. As in all of stable management, safety is first and foremost. Too often one sees horses being shipped with little or no regard to their comfort or well being. Too often, horses are sent on trips with no protection for head and legs, as well as tails and eyes.

Trailering is one of the greatest risks an owner asks his horse to take. Accidents with horse trailers are all too frequent, even if there is no contact with another vehicle or obstacle. Many of these accidents are due to carelessness.

In order to haul as safely as possible you should begin with a good trailer or truck. Single horse trailers are most commonly found in the one-horse stable—one horse for one trailer. While this is a common thought, it is often wise to invest in a two-horse trailer, for reasons of safety as well as expediency, which we will see later on. Single-horse trailers have a tendency to be built too narrow for safety. They also have a tendency to sway when towed under high speeds. This is not to say that single-horse trailers should never be used. If the person doing the hauling is reliable and has had hauling experience, go ahead and use your single-horse trailer. However, hauling should be done by someone who knows what he is doing.

Buying a double-horse trailer, when you only have one horse, may not seem to make much sense. However, if you do a lot of hauling, such as to horse shows, then it makes good sense. By using one side to haul your horse and the other side to haul your gear, such as tack and feed, you eliminate the need for a station wagon or another car along to help out

One of the best types of horse trucks for hauling with the greatest ease, both for horse and driver.

with the transportation of your equipment. By carefully loading hay bales, tack trunks, and saddles, etc., in the unused part of the trailer, you save on time and space at a show. Double-horse trailers can also double as dressing rooms while at the show and can often be made into an overnight stall for your horse, by removing the partition in the center of the trailer. Remember, however, if you are going to use the trailer for equipment and horse hauling together, do not haul pitchforks and shovels, along with the horse. These should be taken in the truck or trunk of the car. These items become dislodged too easily from their moorings and can become deadly weapons in a moving trailer. Also, things being hauled in the extra part of the trailer should be tied down or to each other in order to reduce shifting during stops and starts.

This large, triple trailer has two partitions that may be removed.

Horse vans and trucks are another way of hauling your horse. Trucks often have an advantage over trailers, since they are not as subject to sway and shift, if the horse moves suddenly to catch his balance. Also, they are not as prone to being moved by the wind on gusty days, as are trailers.

Both trucks and trailers should be constructed so that the horse can enter and exit with ease. The animal should not have to jump, lunge, or leap to get in or out of the means of conveyance. Floors of trailers should be padded with either a rubber mat or thick straw. The rubber mat is preferred, since it soaks up shocks from the road better and it can be easily removed and cleaned. It also does not have to be disposed of after the trip has been completed and does not leave bits and pieces of itself all over to mark its path. The rubber mats are also easier for the horse to keep his balance on, since they do not shift or bunch up and make lumps in the trailer, as the straw will do. Neither are they slippery, as is straw.

Straw can be used to bed the trailer down, if used for an overnight stall. Horses should not, if possible, be tied standing in a trailer overnight. If you have no accommodations for your horse and must remain where you are over night, it is better to tie the horse to the outside of the trailer, since standing on the hard, metal floor overnight is bad for the feet. Also, if the trailer does not afford the right amount of room for the horse, he may get cast in the trailer if he should decide to lie down.

This is a good example of a truck that may be used in hauling horses.

This truck has an excellently low ramp for ease of entering.

Only if the trailer converts to a size large enough for the horse to turn around in, at his liberty, should he be kept in a trailer overnight. If he cannot do so then he sould be tethered outside IF there is no other alternative. Standing on ground or grass will not do him any harm, as might standing on hard metal all night.

Regardless of the arrangements, you must realize that this procedure is strictly for those horses that are used to travel, used to being away from their own stalls overnight, and who remain calm and unruffled by any new experience—FOR THE SEASONED VETERAN ONLY.

All trailers should be clearly marked to indicate to other drivers that there are live animals contained within the vehicle and that they should be cautious around your truck or trailer. The warning lettering should be done in florescent paint so that it is easily read at night. An alternative is reflective paint, of the type used to mark highways and road markers. All trucks and trailers should be equipped with running lights for night driving, and they should be in working order at all times. Electric brakes are another must for anyone that cares about his horse's safety and comfort.

Now, what does a horse wear when traveling? Well, he can wear any number of things. A blanket is a must for travel, even in the summer—a summer weight sheet in summer and a heavy blanket in winter. If it sounds absurd to blanket a horse in the middle of July or August, I assure you, it is not. Even the best, most correctly engineered trailers will have

Notice the large lettering across the rear of this trailer. It clearly states that there are horses being conveyed.

drafts when moving. Naturally, the faster you go, the greater the draft becomes. Even in the middle of summer a horse can catch a cold from being too long in a draft, and drafts that blow on chests and backs are the worst. Bronchial colds and kidney trouble may be the result, if you neglect to cover your horse for a trip. Naturally, the colder the weather, the heavier the travel blanket. Too, if you have worked your horse enough to produce a sweat at a show, you don't want him standing around catching cold while he cools off. A blanket tossed over him while he walks not only keeps him from catching cold, but keeps him from stiffening up, as well. Too, a blanket will keep him from marring his coat, should he accidentally rub up against the trailer during the trip or while idle at the show. The blanket is a must.

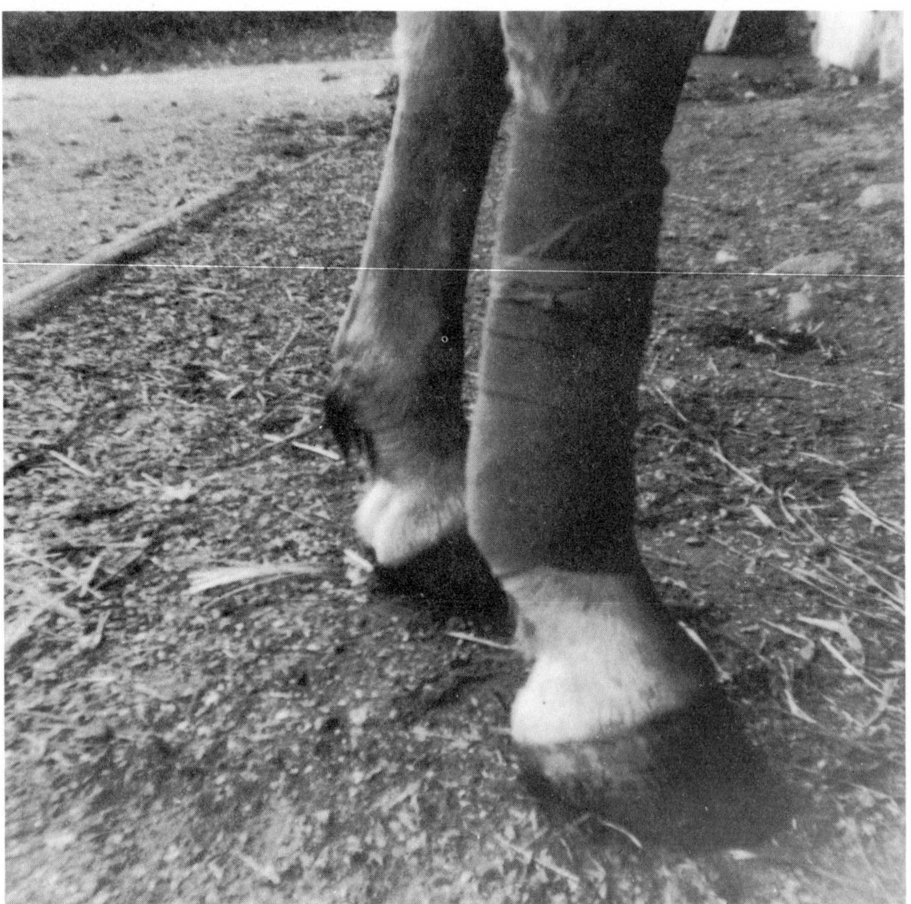

A well wrapped leg, without cotton padding. A leg may be wrapped like this for training sessions, rather than for shipping.

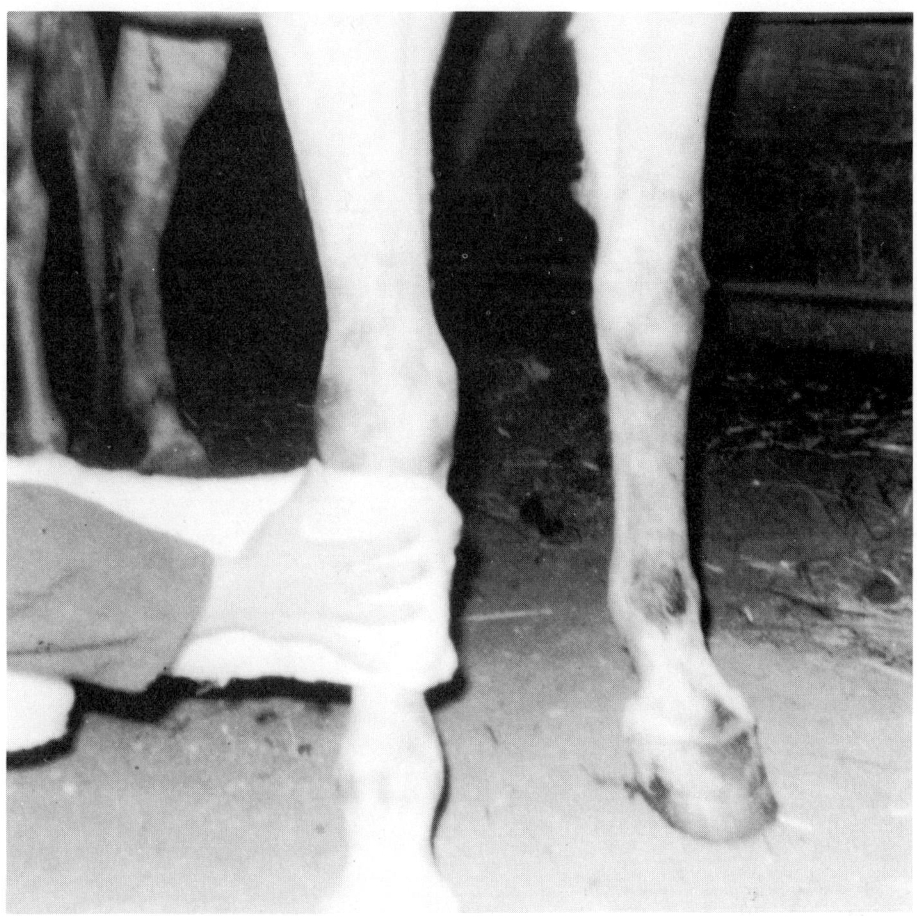

Beginning to wrap a leg for shipping with sheet cotton.

Another must are leg wraps. In order to be the safest, the legs should be wrapped whenever the horse is shipped, whether it be across the state or just across town. A slip, a bump, a cut can cost a lot of money in vet bills and perhaps the use or life of the horse. Horses that are trailered too fast are the ones most commonly seen with cut legs after shipping. Leg wraps can help prevent this.

To properly wrap a leg for shipping you should first start with sheet cotton. This can be bought in the correct sizes and thicknesses at your local tack shop. Regular sterile cotton from the drug store may also be used, but may cost more and has to be pulled or cut apart and thinned to the proper weight or thickness. When you wrap the leg you should first place the cotton around the leg, from just at the knee to the pastern joint.

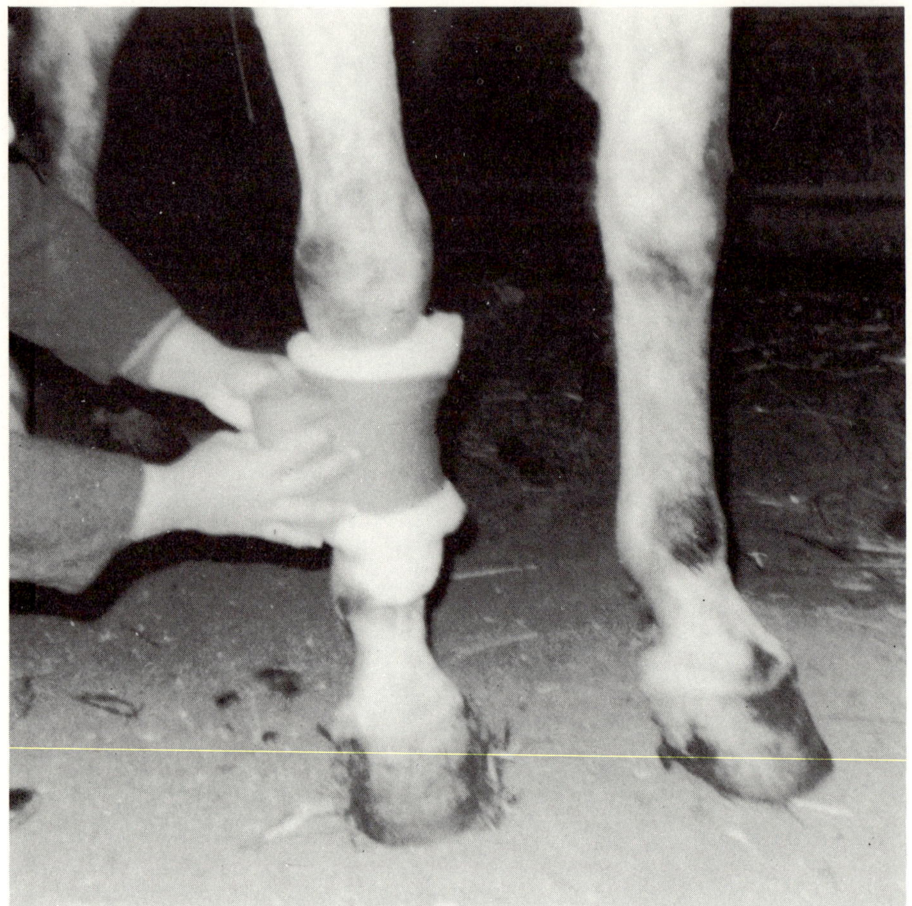

Adding the leg wrap or bandage.

It should be wrapped so that it overlaps, leaving no seam. Next you should take the leg bandage and, beginning at the top, leaving a bit of cotton sticking out to prevent shifting or sliding down, wrap tightly downward. Continue down the leg to the pastern joint. Leave a bit of cotton showing, to keep it from riding up under the bandage. Wrap the bandage back up the leg and tie off, near the middle of the cannon bone.

When wrapping, remember, keep the tension on the bandage as constant as possible, to prevent slipping or loosening. Wrap firmly and smoothly. When you tie off the bandage, tuck the loose ends inside the bandage to keep them from getting caught on anything the horse may walk by or may brush against. Tie the bandage tapes in a single knot and a bow for quick release, if necessary. When you remove the leg wraps, you should roll

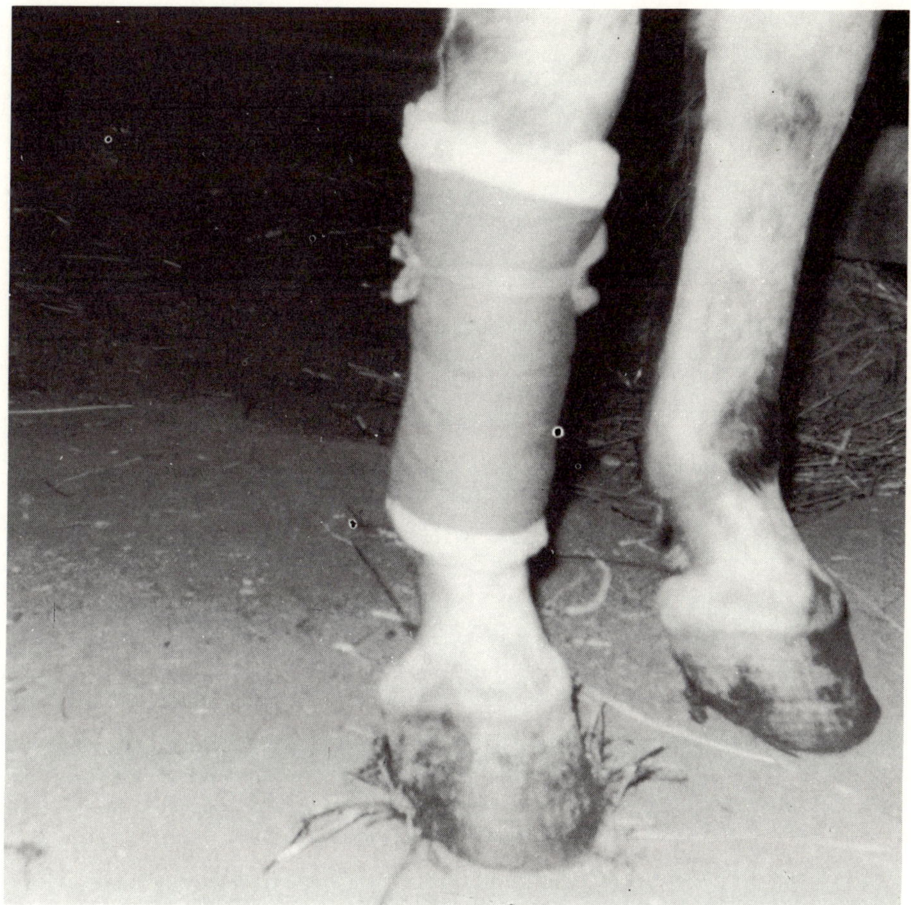

A finished wrapping job. This leg is ready for travel.

them up as you remove them, if they have not become wet or soiled during your trip. This makes them ready to be put back on the horse whenever they are needed or for the return trip home. If the wraps have become wet or soiled, allow them to air dry, if you are away from home, then rewrap them. If you are at home, wash them out after every use. That way they stay soft, pliable, retain their stretch, and are always ready and presentable when needed.

Next is the tail wrap. By right, every horse should have his tail wrapped when traveling. One reason is to keep the hair neat should your horse have a penchant for leaning back and resting on the tailgate of the truck or trailer. Surprisingly, many horses that are used to traveling do this. It seems to be away of taking a rest or getting their weight off their feet

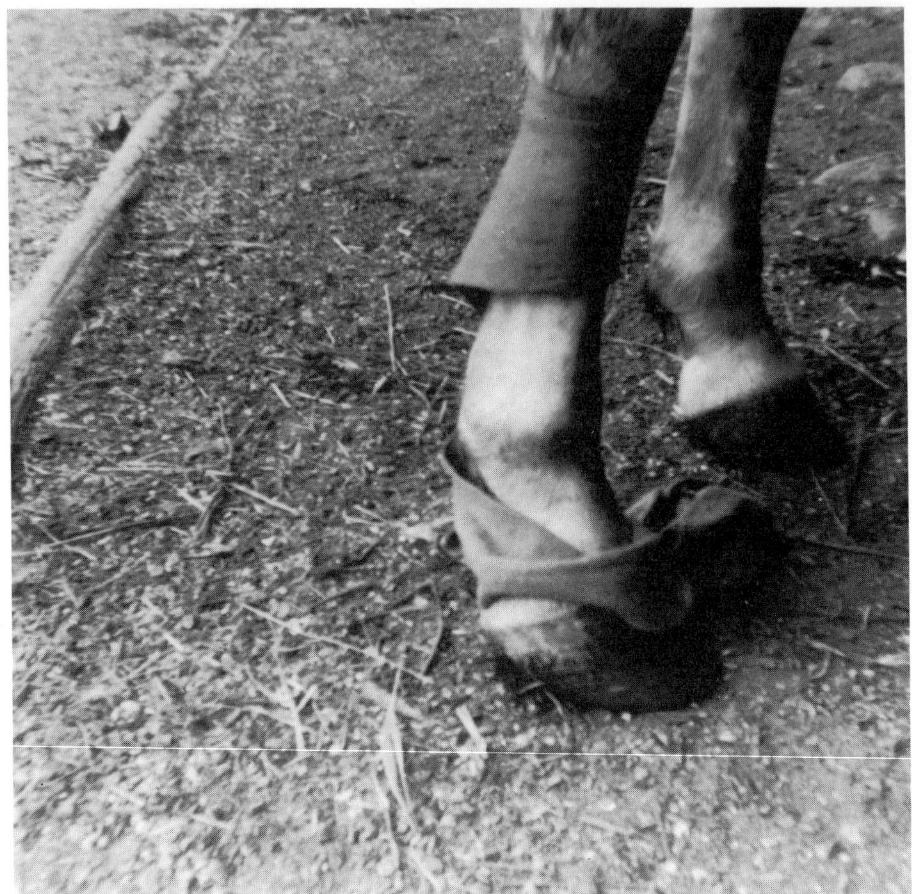

Never take bandages off or leave a horse like this.

when traveling a long distance. If you travel along distance with your horse in the trailer, then you should stop, get him out, and walk him at least once every hour. Traveling, standing up confined in a trailer, is tiring, since it is not natural to the animal. Also, it gives him a chance to limber up and relax. Horses get stiff very quickly in confined spaces.

Another reason to wrap the tail is for protection. While the dock of a horse's tail is not too sensitive when it comes to pulling of the hair during grooming, the dock of the tail is prone to injury when pinched between the rest of the horse and the metal tailgate of a trailer. Fast starts are the major cause of this type of injury to the horse, and tail wrapping can help to prevent this. If the injury is severe enough, to the point that the bone is damaged, it may result in the loss or partial loss of the tail itself.

To properly wrap a tail, start as near the top of the tail as possible,

Beginning to wrap the tail for travel.

placing the end of the bandage on the top of the tail. Do not start on the under side of the tail, since the end of the wrap may rub against the skin and chafe or rub it open. Smooth all of the hair down and begin by passing the roll of wrap under the tail and across the top of the first placed end. Wrap tightly, smoothing the hair down as you go. Wrap down to the end of the tail bone, holding the hair together. Then wrap back up the tail and tie, either in the middle of the tail or about three-quarter of the way up. Do not tie the wrap where you started, since the knot may rub against the horses skin, should the wrap shift for any reason. Make a single knot and bow as you did with the leg wraps and tuck the ends out of the way. Again, roll the bandage up as you remove it in order to make it ready for instant use again.

The next thing that should be worn, but is seldom put on a horse, is one of the best things you can buy for your horse's trailering safety. It is called a head bumper or head guard. Made of leather and lined with felt, the bumpers fit over the horse's poll and protect that most sensitive area from bumps and blows. Bumpers are held in place by the halter crown piece and fit over the ears, leaving them free to be turned and used. Too often a horse will toss his head, either out of boredom or to catch his balance if the trailer is started too fast. In doing so he may easily strike the top of his head on the low ceiling of the trailer. He may do irreparable damage to himself, simply because his owner did not take the time to care about his safety.

If possible, never haul in open-topped trucks. The danger of a horse getting excited and rearing up and falling off or out of the truck is all to clear. If, however, you must, if there is no other alternative, then by all

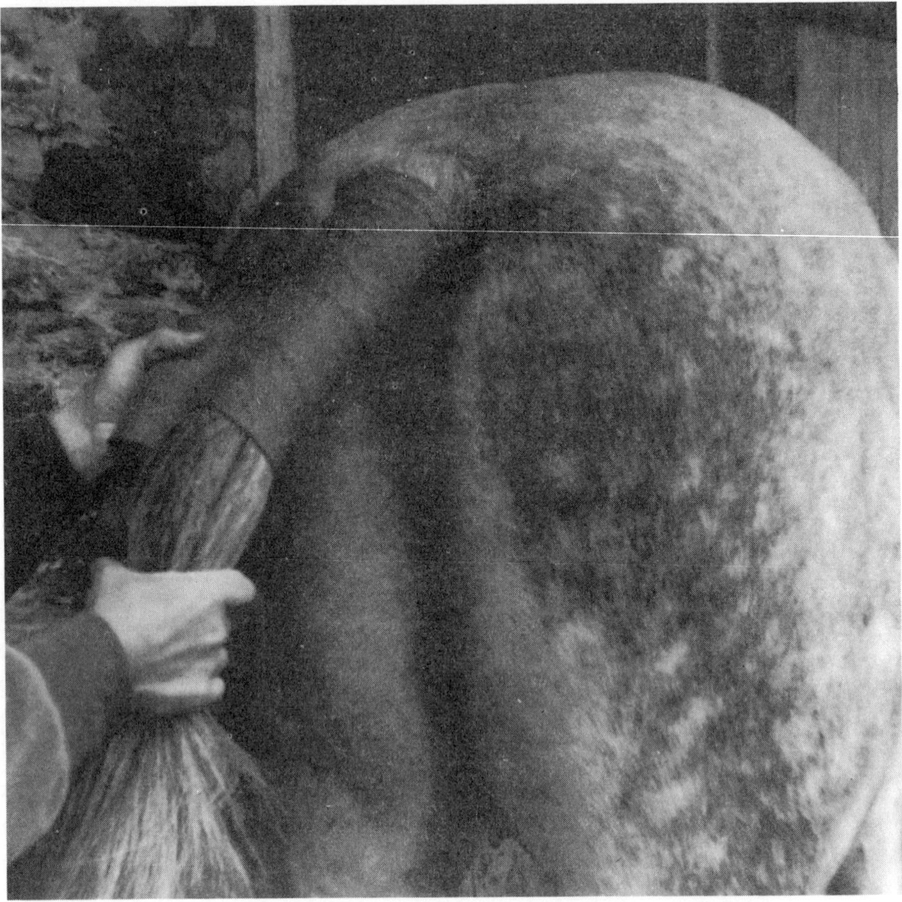

Halfway through the wrapping. Keep the hair held tightly.

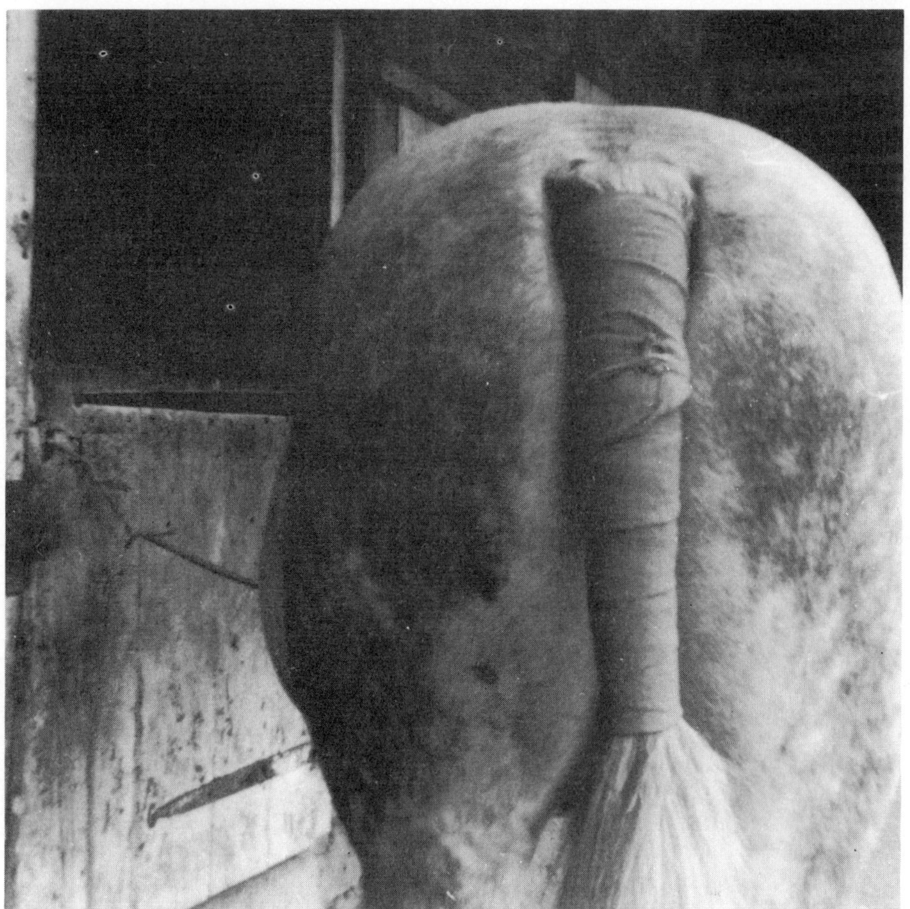

Tail completely wrapped and tied off.

means, have your horse wear goggles. Horse goggles are made of Plexiglas, to prevent shattering, and are padded with rubber around the edges. Trailering in the open is dangerous enough without adding to the danger by permitting the horse to be exposed to all manner of flying dust, dirt, and insects. A piece of dirt that would just produce a blink on a horse that was standing still can blind when striking the eye when a horse is moving at the rate of speed produced by a truck or open vehicle.

Of course, the horse halter and lead shank or trailering tie should be strong, well made, and not frayed or worn badly. Too, your trailer should be free of bad rust spots on the body and floor and the tires must be good enough to pass inspection. When shopping for a trailer or truck, look around and see which ones have the fewest sharp edges inside, the best

A horse properly dressed for travel. Notice the head bumper.

floors, the best axle set up, and the most head room. All trailers and trucks should have adequate night lights and all trailers *should* have electric brakes.

Now, what about speed? How fast should you travel, or how slow? That is entirely up to how used to hauling you are and what kind of a hauler your horse is. Some horses won't tolerate speed, regardless of how many times they are hauled. Many will let you know it by repeatedly kicking the back of the trailer. Many horses that are unused to shipping cannot stand fast trailering, simply because they are not used to the motion. If you hear the clatter of feet quite frequently while on the road, you are traveling too fast for the horse. Unlike humans, most horses do not like speed for speed's sake. Too, horses do not get a thrill out of taking a roller-coaster road at sixty miles per hour for the sensation it produces. Some humans like the feeling, most horses get sick. If, after you have trailered your horse, he is listless and will not eat, more likely than not, he is feeling queasy in his stomach from the ride you just gave him. Since a horse does not have the ability to regurgitate (throw up), the feeling will just have to pass, which can sometimes take more than twenty four hours.

Consider this. When you trailer your horse you are doing so in order to get him from one place to another in the best possible shape and with the least amount of danger to either him or yourself. He is not out for a joy ride. He will not enjoy one. Fast turns, sudden stops, swift starts, and rapid lane changing can all cause terrible accidents when hauling a trailer.

A fine example of a loading chute.

Accidents with horses in trailers are usually tragedies, often fatal and quite heartbreaking.

If you are learning to haul a trailer, do so without the horse in it. Drive your car or truck around town for awhile with the trailer empty, just to get used to how much room you have with it, how well you can park it, and how your car or truck responds when hauling a ton or better of trailer after it. Remember, you will have another half ton to haul when the horse is added. If you are hauling one horse in a two horse trailer or larger, put the horse in the front of the trailer and behind the driver. The trailer is most easily controlled with the horse there. When loading, use loading ramps or chutes, whenever possible. They will keep the horse in the proper alignment with the trailer entrance and lessen the chance of him refusing entry or swerving away.

If you think enough of your horse to trailer him to shows, better quarters, or for any reason, then you should think enough of him to give him the best protection possible.

11

BITS AND PIECES

ONE THING TO KEEP IN MIND when running any stable is to be sure the barn itself is kept up and in good repair. Broken windows should be fixed immediately and not allowed to go unattended. The danger of the horse getting badly cut by the broken glass is quite real and the constant draft blowing into the barn is an invitation to pneumonia or severe colds, not only for the horse in the stall with the broken window, but to the whole barn.

Another very serious problem arises when boards between stalls get kicked or broken out and leave large, open gaps through which one horse can easily gain access to another. These gaps are invitations to biting, kicking, and fighting matches. Also, a curious horse may try to enter the other horse's stall by way of the hole and get cut, impaled, or cast between stalls. Also, if any lower boards remain below the gap, the horse on either side may break a leg, should he accidentally step through the hole and panic.

Good maintenance is a must around any stable.

A good idea, around any stable, is to have access to water in any and all paddocks or corrals. One good way to water is a spring fed trough. The water flows through a pipe, from the spring house, and flows out again by way of an overflow pipe. Here the horse can freely drink of constantly fresh water, water that is never still or stale. A large goldfish or two in the trough will keep down all insects that may collect near the trough and the horse won't mind one bit.

Another good watering method is an old bathtub. Filled by a hose that is fed into the tub through the fixture holes, it can easily be filled,

Broken out window panes, like this, permit drafts throughout the barn and should be fixed as soon as possible.

simply by turning on the faucet to which the hose is connected. The drain may be opened for cleaning and draining, simply by pulling the stopper. Tub stoppers are available at hardware stores. Though not as good as the spring fed trough, the tub does its duty well if emptied and cleaned each day, before and after use.

What do you do if you are a very short person trying to get on a very large horse? Or if you are a novice and unsure of yourself? Or if you are past middle age and have lost some of your spring? The answer: build yourself a mounting block. The mounting block or platform provides a stable, safe, and secure platform that gives you added height and enables you to mount a horse more easily and gracefully than from the ground.

The best type is the three step, which is bolted together for safety. Oak boards for the steps and supports give added durability and take weathering well. The wide-top step makes a secure platform from which to mount.

How does one keep small articles for grooming, extra lead ropes, etc. safe, yet ready for use? The answer: in an overseas chest that has a removable shelf. This keeps all small articles neat, contained, and clean until used. Medicine and bandages, too, may be kept in chests if there are no medicine cabinets available, or if there is no room for a wall chest. The chests are quite easily built by anyone who can handle a saw and hammer.

If you enjoy jumping your horse you might be interested in building a show-jumping course of your own.

Broken out boards between stalls should be repaired **IMMEDIATELY**.

An excellent example of a spring fed watering trough.

Stable management, though work, is enjoyable when done correctly. New and better ideas are being conceived all the time, both for master and horse.

If you are a master, be a good one.

If you are a manager, be the best.

An overflow pipe carries off excess water from the trough.

This type of bathtub is one of the best for watering horses.

A fine, sturdy mounting block.

Notice the wide top step for mounting. Also, the space between the block and the fence should only be wide enough to permit the horse to pass between without brushing either.

A fine example of a jump course that you might want to build.

Medicines, too, may be kept in trunks if you have no medicine cabinet.

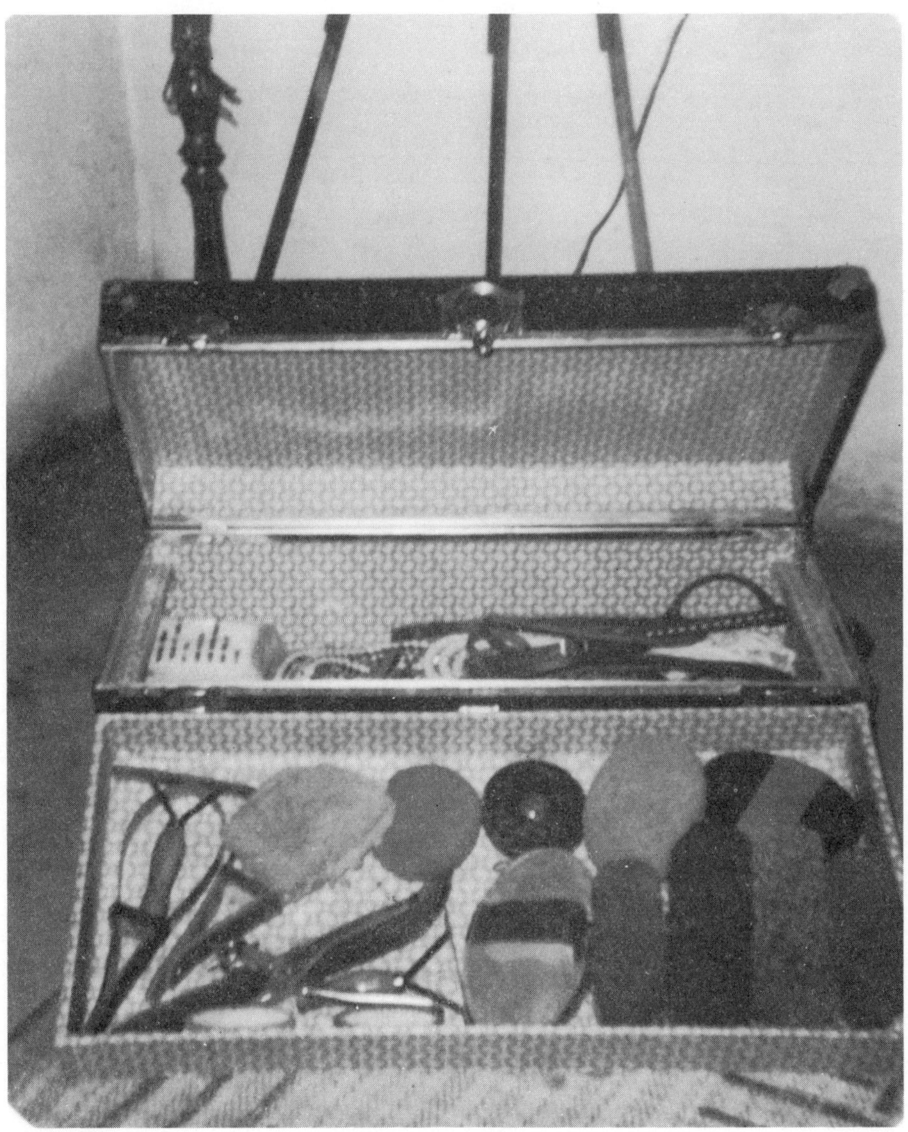

A good method of storing brushes, etc. if you don't have room for a tack or brush locker.

A detail of one type of jump found on most jumping courses.

INDEX